FLYING AIR ELECTRIC
a novel

By
JONATHON GLANE

Royel Company, LLC
ROYEL WRITERS
Fort Lauderdale
royelco@hotmail.com

Dedicated to the memory of 'Bugs'
my brother who left us long before his time.
There never was a better childhood storyteller.

ACKNOWLEDGEMENTS

The publication team at Booksurge provided more than a significant amount of help producing this novel.

Stephanie Robinson-Crane and the TDF1 team; Jenny, Lauren, and Amanda along with Abby Harris and the DC3 team; Blair and Emma have my everlasting gratitude.

Contents

Chapter One: WHAT ABOUT ME?

Hector looked at the couple sleeping peacefully. He could not understand how they could be doing this when he had been wide-awake and hungry for some time. Hector decided to do something about it. He walked the entire length of Brad's body before jumping onto Lissy's softer frame for the same stomping excursion.

"I think your cat is hungry," Lissy murmured.

"He's up for adoption," Brad groaned sleepily.

"I'll think about that, but in the meantime his owner gets to get up and feed the poor fellow."

Lissy watched Brad as he moaned and groaned in the process of slipping on his pajama-bottoms and getting ready to feed Hector, who was now rubbing against his leg. Lissy had always admired Brad's athletic body that belied his age of forty-six years. She liked his auburn hair that was only slightly darker than her own with its ever-so-slight tinge of gray around the temples. She liked his boyish face too, with that lopsided grin that seemed always to be at the ready. Of course, she liked Brad enough to spend most of her free time with him and to treat his apartment as her own much of the time. After all, he had asked her to marry him and this gave her a realistic indication of the affection he felt. She was not sure if her feelings for him were strong enough to accept his proposal, so she told him she would think about it and she was still thinking after all these months.

Hector was getting impatient. Twice now, he had started to run ahead of Brad into the kitchen and twice now, he had stopped as Brad changed directions. Now Brad was opening the drapes. Surely this could not be as important as the growling complaints Hector's stomach continued making. Finally Brad began moving in the right direction and Hector was quick to trot ahead to his feeding area in the kitchen holding his tail, as he did, at the vertical with a slight crook at the end.

Lissy liked it when she could stay in bed with the light of day coming through the window. She watched the bird feeder outside that Brad had hung on a branch and she admired its frequent visitors. There they would come by ones or twos to peck away at the seed the feeder provided until the next visitor would threaten a painful jab if they lingered longer than a certain amount of time. Then the new guest would take a turn until the cycle was repeated and a new visitor would get a chance at a snack. Much of the time size mattered in the pecking order, but often it did not. The smallest visitors seemed capable of sending the largest ones flying if enough territorial aggression was displayed.

In only a short while, Hector rejoined Lissy and assumed an Egyptian lion's posture on her legs. He had been as finicky as possible with the Fancy Feast Brad set out for him, and now he was more interested in the view Lissy was enjoying outside the window. Hector's tail was twitching in anticipation as he imagined the giant leap he would make to capture the bright red bird Brad called 'Mr. C.' the cardinal. His master was sure to be pleased when Hector brought his quarry to the front doorstep as a feast they both could share.

Lissy was so caught up in the peaceful tranquility of the exhibition outside; she failed to notice Brad leaning against the doorframe admiring her. He marveled at the beauty he beheld. If Lissy had been wearing makeup, she would have looked different, but she would not have looked better. Her auburn hair, shining with a natural luster and highlighted by scarlet 'hair pretties', was piling on her shoulders in just the right amount. Her brown eyes were twinkling while her full mouth was hinting at the enjoyment she was feeling about the spectacle outside. The rumpled pajama top of Brad's she wore concealed the upper part of the shapely and completely feminine body he knew to be lounging beneath the covers.

Brad was experiencing a little guilt about being with this lovely creature who was twelve years his junior when Lissy, realizing he was there, looked at him smiling.

"Is breakfast ready yet?"

There would be time on this Saturday morning to enjoy Brad's version of bacon, eggs, hash browns, orange juice, and coffee before Lissy would help Brad straighten up the place and get ready to go to work at her regular job at Grubby Ernie's Tavern.

Lissy Edington had worked at Grubby's for seven years and she was getting a little tired of the place. She knew that being a bartender and cocktail server was not something she wanted to do for the rest of her life. She also knew that she could quit and move in with Brad anytime she wanted, but her independence would not allow her to do that. Besides, over the years at Grubby's she had built a clientele that had become so regular they were almost like family. Turning her back to them would be a very difficult thing for Lissy to do. Neither could she discount the substantial amount of money provided by the job at Grubby's.

Tending bar was not Lissy's only source of income. She and Brad and her father, Captain Ed, had developed a crab trapping business that was operating successfully. Utilizing a productive field off the coast of New Jersey, Captain Ed had responsibility for the day-to-day operations while Lissy and Brad assumed the passive role of investors. Their investment was paying off, but it would be a number of years before it would amortize. In the meantime, the income was only slightly greater than the expenses, so this was hardly the enterprise Lissy could fall back on if she left

Grubby's. So, even though it was Saturday and a beautiful day at that, Lissy prepared to go to work by 11:00 o'clock and get ready for the noontime crowd.

Brad Ganderson had no such obligation. His five-day job gave him weekends free. Much of his leisure time was spent surfing the web and checking his email when he was not spending time at Grubby's just to be with Lissy.

Brad's regular job was to manage the operations of Powervert LLC, an entrepreneurial organization that produced income through the technology of transferring electrical power wirelessly. Powervert's customers, electric utility companies, found it economically beneficial to use Powervert's 'power-beaming' services in places where it was impractical or overly expensive to transmit electricity with wires the conventional way.

The latest projects Brad and Powervert had undertaken were partnering with a number of microwave energy companies, so the stations they had already established to send microwave signals from hilltop to hilltop could be modified to send electricity also.

Brad's coworkers were also his partners. They had formed Powervert into a Limited Liability Company with Rena Offlin as Managing Member and Jason Grilling along with Brad as General Members. Rena managed the day-to-day business of the company while Jason, with a team of three scientists, directed the Research and Development department.

Recently, Jason had informed Brad and Rena that the research team had undertaken a promising and exciting project. They were exploring the ways their electricity transmitting equipment could be used to send power to unmanned aircraft used for surveillance. Jason believed, and the U.S. Military agreed, that this type of aircraft could monitor the activities of terrorists and other clandestine organizations, and in that way help bring these threats to world peace under control.

Until now, the most reliable vehicles for continuous observations were satellite-based and they left much to be desired. Placing a satellite in a geosynchronous orbit would accomplish the task, but this method required a different satellite for each geographic area under surveillance. Besides, the images captured at these vast distances produced a resolution that resulted in a picture of poor quality.

On the other hand, electrically powered aircraft operating wirelessly could return sharp resolution images and remain aloft for days or weeks, or as long as the electrical supply lasted. Also they could be shifted easily from place to place and any number of them could be launched according to the situation at hand. And, as a bonus, they would have all the electrical power needed to operate radars or cameras, and to send detailed information back to their bases.

Brad would use much of this Saturday off work researching 'Electricity Powered Aircraft' and other topics on the Internet. The demand for the technology was certainly in place and once it became available, it would be Brad's job to organize the elements into a package that could be sold to the military or other governmental organizations.

Brad would not be surfing alone. Hector joined him as soon as he realized an unoccupied lap was available. Brad did not mind the company. He was able to bounce his ideas off Hector and he knew the answers returned would never fail to be the correct ones. Hector's eyes would be tightly shut while his head was erect and the smallest purring sound would be coming from deep inside. This posture along with sharp claws gently kneading Brad's leg gave positive reinforcement to each and every one of Brad's suggestions.

Brad told Hector the simplest way to supply electrical power to unmanned aircraft would be to provide each one with a ground-based generator. Hector maintained his posture, but he made no sound and his claws remained sheathed.

Brad decided that might not be such a good idea since power-beaming required a line-of-sight to the aircraft; a generator might be stationed too far away if the area of surveillance happened to be deep inside enemy territory. Hector's claws dug gently into Brad's knee.

Perhaps if the power beam was bounced off a satellite in geosynchronous orbit, the receiving surveillance aircraft could be in any number of locations whether they were near or far from enemy borders. This suggestion prompted a purring and a kneading.

Jason and his team of scientists would have to be consulted in order to verify whether the power-beaming equipment was capable of sending electricity the distances needed to beam up to a satellite and back down to earth. Brad utilized the touch-screen on his iPhone to dial the pre-programmed numbers of Jason and Rena.

"What say we all meet at Grubby's around 5:00 o'clock for drinks and appetizers?" Then he apologized to Hector. "Sorry, this is for clawless and purr-less folks only."

Grubby's was alive with activity. All the sports fanatics were there to watch the action on the wide screen HD televisions scattered around the room. Lissy was as busy as she could be even with help of the extra bartender who had been called in for the occasion. Brad was not surprised that his regular table near the end of the bar was unoccupied. After all, it was located in the least advantageous place for watching TV, but Brad liked it because its position allowed him to talk to Lissy whether she was behind the bar or waiting tables—as she was this day.

Only a short while after Brad was seated, Rena and Jason, a strikingly handsome couple, walked through the doorway. The bright floral print on Rena's white dress contrasted her clothing the same way her bright red lipstick accented her light complexion. When she re-set her sunglasses atop her wavy blond hair, her blue eyes framed with black mascara completed the Princess Diana-like portrait of beauty. The high heels Rena wore, elevated her to quite-near Jason's height. Only a little shorter than Brad, Jason's appearance complimented Rena's glow with his dark brown hair and dark complexion. Handsome in his own right, the vision of the couple entering the room gave the impression of sophistication and elegance.

"Good to see you," Brad offered as he arose in greeting before exchanging embraces with Rena and shaking Jason's hand.

"Same here," Rena responded while Jason was saying, "You too," before continuing with, "Wow, this place is hopping."

No sooner were they all seated than Lissy walked over, hugged Rena, and said, "Hi, Jason. I see you guys don't get to see enough of each other during the week. What can I bring you to drink?"

"I'll have a Margarita," Rena answered.

"Me too", Jason echoed. "Make mine on the rocks."

"My beer is fine for now," Brad ascertained to complete the order.

While Lissy was away getting their drinks, Brad asked the others if they were ready for an appetizer. They both agreed that they would prefer to wait a little while before ordering food.

"You guys look too pretty, and this is hardly the place for business, but here goes anyway," Brad began. "I've been thinking about the different ways to send power to unmanned drone aircraft and it seems that bouncing an electrical current off a satellite makes the most sense. Tell me Jason, do you think that our system has enough oomph to send an electrical beam that far?"

"Where we are right now, the answer is no," Jason answered. Then he continued, "Let me rephrase that. Our system has a range that would reach up to a satellite and back, but the problem is the power loss that comes from sending it that far. Unless we can find a way to concentrate the beam enough to minimize the loss, the electrical current reaching the drone aircraft would not be enough to make it operate."

Lissy brought the Margaritas and Brad said, "I'm ready now."

"I'm way ahead of you," Lissy countered. "Here's your next Coors. I decided to bring you another anyway and no, it's not on me!"

"You go girl," Rena encouraged as Lissy left. Then she added, "I know Jason's group needs to bring the technology on line to make this all happen and I'm confident that once the method is available, Brad will put the system together. What neither of you know is that we could be barking up the wrong tree." Brad and Jason exchanged questioning glances before Rena went on. "You both know about the meetings I have been having with the boys at the Department of Defense. Well, during our last meeting they told me about a new technology that some companies are working on. It's called nanotechnology."

"Sure, making things super-small has been around for a while," Jason interjected.

"That is not where I'm coming from," Rena clarified. "I'm suggesting that nanotechnology can be utilized to provide the parts needed to create a drone so small it would be virtually invisible. But that's not all. I believe it's possible to set up a chain of these mini-drones in a way that they could be used as repeater stations to transfer wireless electricity wherever it's needed and no one would know they were even there."

"Move over, Jason, you've been trumped," Brad teased.

"See why I love this girl," Jason said. "It's a contest between beauty and brains and the brains are working overtime, but they can never win."

Rena's attempt at being coy only resulted in her looking more radiant than ever.

Brad wanted to find out if this sort of technology would fit with Jason's development plans. He thought it was important for the three of them to be on the same page with Powervert's goals in mind. If a new technological direction was to be taken, areas of responsibility would need to be determined. Clarification was required.

"I have all the faith in the world in our research department and its capability to stay on the cutting edge of things," Brad said. "The *caveat* here seems to be the number of different things needed to bring this nanotechnology thing about. If we re-assign part of our research team to the project of developing mini-drones, it will probably hurt our chances of staying dominant in the power-beaming field."

"I believe that's right," Jason agreed.

Rena agreed too before suggesting, "If we decide to go along with miniaturizing our power-beaming method, it might be a good idea for me to get back to the guys at Defense and find out who's the front-runner at developing small, unmanned drones."

Brad added, "And if some company is advanced enough and is willing to work with us, we can concentrate on the things we do best."

"You guys are talking some serious business," Lissy observed as she joined the group during a short break in her duties.

Rena and Jason had no problem accepting Lissy into their inner circle. She had not known Jason that long, but she was long time buddies with Brad and Rena since they had been scuba diving together several times and Lissy had visited Powervert's offices occasionally.

Brad put his arm around Lissy's waist as a sign that she was considered a member of the group. They both listened as the business discussion continued.

Jason was saying, "I think it would be relatively easy to make our power-beaming equipment smaller and still be able to send enough juice to make the drones work, especially if the drones themselves are miniaturized."

"OK then," Rena concluded, "I'll check on the state of the art for mini-drones while Jason researches making the power-beaming stuff smaller and Brad continues working out at way to tie all the elements together."

The Listener removed the headset and walked away from the machine. The voice-activated recorder had condensed the recording time, so the rest of the conversation could be listened to easily at another time. The Listener had heard Powervert's new plans, and immediate action needed to be taken.

Chapter Two: EVESDROPPER

Rena's meeting with the 'boys' at Defense was fruitful. There was indeed a company, not far away, that had developed a palm-sized helicopter using carbon nanotubes for the structure and a polypropylene sheath for the covering. The translucence provided by the plastic that covered the drone made it almost invisible, even at close range. And when light blue paint was applied to the interior parts—the motor, the camera, and guidance systems—it became even less likely that it would be noticed while flying overhead.

Other than its near invisibility, it resembled a Bell Helicopter except for its wider rotors, which departed from the norm with their broad and de-curved blades.

"This isn't the only company out there working on Unmanned Airborne Vehicles using micro-technology, but these people seem to be far ahead of the others, and they have indicated a willingness to talk with us," Rena explained to Brad and Jason as soon as she could get them together after her meeting with Defense.

"So, assuming we can strike up a deal," Brad said, "Jason will have his work cut out for him to develop a way for these babies to receive and send electrical current."

"And Brad will have his work cut out also developing a network that can get the electricity to the drones," Jason added.

"Well OK then," Rena went on, "what do you guys think? Can these things be worked out?"

"I'll need a big map," Brad joked.

"I'll need a huge magnifying glass," Jason joined in the fun.

Then Rena added, "And I'll need a larger-than-life level of tolerance to put up with the two of you!"

After the meeting, Jason assembled his research team to begin the daunting task of miniaturizing all the power-beaming technological equipment they had developed so far. One of the scientists jokingly suggested they put the parts in a copier and punch the size-reduction button.

Jason added to the amusement by saying, "The only way that could work is if you use a 3D copier."

And as strange as it sounded, that was exactly what the scientists did except the 3D copier was actually AutoCAD; the proprietary computer drafting software they purchased from Autodesk, Inc. They were able to shrink the size of each of the parts so it would fit into the 'palm-sized' drone, and still be able to function as originally intended. The last challenging operation was to construct the newly re-sized elements, and then assemble them into a functioning arrangement.

Rena spent much of her time with Dronomics, the developer of the micro-sized Unmanned Airborne Vehicle or UAV. She wasted a lot of time by telling Dronomics that their product was not exactly what Powervert was looking for, but they would settle for it. The director of marketing did not buy into any of that, and refused to sell the product unless a royalty payment could be agreed upon. Rena declared that the ultimate use of the micro-UAV was not such that a royalty agreement could be reached, and she attempted once more to buy the product outright. After a great deal of old fashioned haggling, they finally agreed to a joint-venture where Dronomics and Powervert would share equally the profits gained from the sale of the finished product, and the costs associated with marketing would be split between them.

Likewise, Jason and his team were making progress, but it was developing much slower than expected. They soon found that miniaturizing a part using a computer program was one thing and actually building it was quite another. Most of the components they needed were simply not available on the open market, and each one they had to build from scratch required time-consuming microscopic construction and assembly.

Slade, Gordon, and Bratley, Powervert's scientific team, were not used to doing their work in miniature. Jason had warned them that mistakes made under the microscope were often magnified. "No pun intended," he would say. But the team believed that 'exponential' described the conundrum much better because each seemingly minor misstep resulted in a major setback.

The lack of significant progress was not the worst aspect of the research. The team members were beginning to blame one another for this or that failure, and Jason was fearful that the entire project could be destroyed if the infighting was allowed to continue. He summoned Slade, the senior member of the scientific team, into his office.

"Don't think that I don't understand what's going on," he began. "Bratley isn't sure he wants to continue with the miniaturizing work, Gordon thinks he's underpaid for what he does, and yet he wants us to go out and buy all new equipment, and you have been complaining about the tight schedule.

"As the team leader, you should be able to delegate responsibilities to Bratley and Gordon to make sure our goals are reached. And they are reachable with the equipment we have. As for Bratley, he has to understand that research and development is the same whether it's big or small and the only thing that needs adjusting is our attitudes toward our projects."

Slade did not waste any time responding. "Saying not to be frustrated is easy … making things impossibly small is not so easy."

"I understand," Jason said. "But the hardest part … drawing the roadmap to take us where we want to go … has already been done. All we have to do now is take the journey no matter how many obstacles we come across."

Earlier in the day, another meeting attended by Powervert's officers, had given each of them an opportunity to bring to light the problems and the progresses they were making.

Rena's project seemed to be under control and on track, but Jason's was proving to be difficult. Brad had always thought that his responsibilities were the easiest and the least challenging. He was right. All he had to do was find an electrical outlet somewhere and plug into it, or provide a generator to do the work.

Later that afternoon at Grubby's, Lissy listened intently as Brad explained to her all that was going on. During her visits to Brad's table between her duties to the other customers, Lissy took interest in hearing all the details about the way things were going for him at work. Lissy thought that was what a devoted mate was supposed to do; listen to their significant other, and she and Brad seemed to be getting closer with the passing of each day.

Lissy was not the only one listening.

A tiny microphone stuck underneath the table where Brad always sat recorded every word they said. The Listener smiled at the clarity of the conversation being picked up by the microphone. and frowned at the problems Powervert was having. There was no need to worry. Progress was slow, but it was being made. All good things come in their own time.

Chapter Three: THE LARGE INSECT

"It looks like a moth," Brad exclaimed when Rena brought the first Dronomics prototype to Powervert's laboratory, "a fat moth with a helicopter's tail-end."

"That's it," Jason agreed. "It's a moth that's a robot. From now on we should call it a 'MothBot'."

Jason had been using schematic plans to construct the elements meant to fit inside the drone, but this was the first time he saw the Dronomics UAV assembled and ready to fly. It only lacked a means of power and Jason's battery-powered electric motor was ready to be installed and tested.

"Does it live up to everyone's expectations?" Rena wondered.

"It's awesome," Slade observed speaking for the scientific team, "but it looks fragile. Are you sure it won't break if I touch it?"

"Let me put it this way," Rena answered. "Nanotubes … the structural material … are said to be 1,000 times stronger than steel. As for the outside skin, did you ever try to open one of those packages wrapped in plastic?"

Smiles brightened the faces all around. They knew how strong plastic packaging could be.

Rena went on. "The body hinges open right behind the main rotor so the insides can be assembled easily. And this black box is where all the controls are."

The pilot's controller she held was about the size of a cigar box. It was painted black and had a joystick and some gauges along with a q-w-e-r-t-y keyboard. A video monitor the same size as the box became its lid when it was folded down.

"The joystick is used to control the … er … MothBot when it is in sight. When it is out of sight, which happens pretty fast, other inputs guide it to certain coordinates using its GPS, or it can be told what to do by simply typing instructions on the keyboard."

"Can I see it?" Brad asked.

"Sure," Rena said placing the MothBot in his hand.

Brad was impressed. He held the palm of his other hand over the small object. It was about the same length as the distance from the end of his fingers to the small part of his wrist and its rotors were as wide as he could spread his fingers apart.

"I'd say it's about eight inches!" Brad exclaimed.

"It's actually 200 millimeters," Jason corrected.

"Picky, picky," Brad accused before adding, "I want to see it fly."

"I'm sure we all do," Rena guessed. "What do you say, Jason? When can it be arranged?"

"Give us two days," Jason responded. "We'll rig it with a wafer battery or two. That should provide enough power for several minutes of flight. But who is going to be the pilot?"

Rena agreed to have someone from Dronomics come over for the piloting duties.

"I can't wait," Brad said.

Derek Sorsky, Dronomics VP and the drone pilot, went to Powervert's offices as soon as he could after Jason told Brad and Rena the MothBot was ready for its test flight. Along with Slade, Gordon, and Bratley, they assembled in the warehouse and sat patiently in steel folding chairs while Derek got things ready.

Derek put on a pair of goggles and then with one unexpected motion of his hand, he tossed the MothBot into the air. The thought flashed across Brad's mind to run and catch it before he saw it simply hanging there in space virtually motionless. Then it began to slowly descend. It was moving downward at a fairly fast pace before the rotor was seen turning.

'Amazing aerobatic dexterity' might have been used to describe the way the MothBot moved through the air. Like some curious airborne creature, it flew to within a few feet of each of the faces staring at it in wonderment. One at a time, the MothBot would pause in front of a person's face, appear to be studying it, then turn and move to the next one. No one said anything. No one could have said anything. The magic taking place before their eyes seemed to be beyond description.

Finally Gordon, scientist No. 2 said, "I've got it. It isn't really there. It's only a hologram!"

The observation brought the others out of their state of amazement.

Slade remarked, "If it is, how is it possible to hold it in your hand?"

Derek laughed but otherwise ignored the question.

"I have to keep it close," he said. "It goes out of sight quickly."

The words were no sooner spoken than the MothBot disappeared from sight. Everyone was wondering where it went when the slightest whirring noise could be heard overhead and there it was, a few feet above their heads.

"Now you see why we have to have more than one way to control it," Derek was saying, "I didn't know where it went either, but I could see where it was going." Derek knew they did not understand what he meant so he clarified it.

"These glasses I'm wearing are actually miniature TV monitors. A small screen inside each lens allows me to see the same things the drone's camera sees and the same thing you can see on the black box monitor. This way I can maneuver the drone the same way a pilot would who was sitting inside the cockpit."

Gasps of disbelief could be heard from every quarter until Derek proved what he was saying by allowing each of the spectators to look through the goggles for a moment. They discovered they could see their surroundings through the lenses as well as the view through the tiny TV screens.

Then Derek continued. "When I don't know where it is, I set its GPS location and altitude for it to go where I want it to go. When I know where it is, I simply use the joystick to fly it where I want it to go.

"Now I'll show you one of the main features of its flying capabilities." Derek moved the joystick and the MothBot responded by flying near the warehouse ceiling and a few feet in front of the group. Then the rotor slowed almost to a stop and the craft seemed to hang motionlessly in space. It was not going forward or backward. It was not a blimp. How was it possible that it could be doing this?

It was difficult to see, but the MothBot was slowly losing altitude while it seemed to remain stationary.

"The craft is so light that air molecules support it much more than they would a heavier object," Derek explained. "We're in a confined space here, but outside at a high altitude, it could take 30 minutes or longer for it to descend all the way to the ground. And now, the controller will be disengaged and we can see what will happen."

Derek set the black box aside and the MothBot slowly descended all the way to the floor. The rotor blades were only turning slowly when the craft struck the floor. Its flexibility allowed its parts to bend without breaking as it hit.

It was Bratley's turn to make a contribution to the conversation. "It would seem that all the years I spent in college were wasted. The scientific education I paid so much money for is telling me the things I just saw are not possible."

"I agree," Gordon chimed in.

"Me too," Slade said. "The things we saw can only be done if the drone is lighter-than-air."

"That's right," Derek concurred. "You might have noticed when you were installing the electric motor and batteries that the plastic skin covering the drone is actually a double layer. I know it's hard to see because the craft is so small, but there is an inner and an outer skin that holds its shape with membrane spacers and the interior of the skin is filled with hydrogen."

"That's why it looks so fat!" Brad exclaimed.

"Yes," Derek continued, "and the rotors are also filled with hydrogen. The result is that the drone would fly away like a child's balloon if not for the weight of the equipment inside."

"We didn't notice that when we were installing the motor and batteries," Slade challenged.

"You didn't notice because we always attach a thin metal weight to the bottom of the drone after it has been charged with hydrogen. That's the only way we can keep them from escaping. The metal weight also holds the drones in place when they are not flying. What we call our 'landing field' is nothing more than a sheet of magnetized metal that the drone's metal weight causes it to stick to like glue."

Derek brought the demonstration to a conclusion by explaining, "Our goal has always been to offset the weight of the equipment with the weightlessness of the hydrogen filled structure to the point where they are almost in balance with just enough positive weight to allow controllability.

"Also, the hydrogen filled rotors and outer skin provide an additional element that the government finds attractive. Should the drone be in danger of falling into enemy hands, a device we call a 'piezo poison pill' is activated. This causes a spark to ignite the hydrogen gas and the entire drone is destroyed in a burst of flames."

With the demonstration over, Rena, Brad, and Jason brainstormed in Powervert's conference room after they had thanked Derek and bid him goodbye.

"By the time we add all our equipment," Jason began, "much of the MothBot's maneuverability will be lost. After all, it's the near weightlessness that makes it fly around the way it does."

Then Jason shuddered at Brad's suggestion.

"Maybe we should try to make the components even smaller." Brad quickly thought to withdraw his suggestion as soon as he saw the scowl on Jason's face. More miniaturization was clearly not an option.

"This idea might work," Rena proposed. "If the operating components can be hermetically sealed, the rest of the inside of the MothBot could be filled with hydrogen."

"That's it," Brad agreed. "Whatever weight is added, can be offset when more gas is added also."

Jason told the others he thought the plan to keep the MothBot as light as possible could work, but he questioned the practicality of something so light in weight a strong wind could easily blow it away. Naturally, wind considerations were not a problem inside Powervert's warehouse, but in the outside world, a stiff breeze could make controlling the drone difficult.

The three officers decided that undertaking solutions to wind problems should be a project for Jason to consider in the future. For now, they would work with what they had and plan on launching the MothBot only on days when the weather forecast called for calm winds.

When Jason excused himself to return to his duties in the laboratory, Rena and Brad had a chance to discuss Powervert's progress privately. They agreed the responsibilities each of them had were at or near fulfillment, and that it was Jason and his team of scientists who were holding up the parade. They made it a point to declare that this was not meant to be detrimental in any way. It was only the way things were although they wondered if Jason was somehow losing his edge under the real or imagined pressure he felt. There was a great deal of pressure on Jason to be sure, but Slade, Gordon, and Bratley were doing the bulk of the work. All Jason had to do was supervise their activities.

The temperature in the conference room must have been hotter than Brad realized because all at once he noticed Rena was unbuttoning things and adjusting things and loosening her hair and things. She must have been dispersing a chemical in the air at the same time because Brad began to be aroused with her sensuousness. Rena was a beautiful woman who possessed more than enough attributes to put her in the 'lustfully desirable' category, but she had never shown him this side of her personality before and he had no idea how to deal with it.

He did nothing.

A chasm of silence began growing between them.

Brad was watching Rena, but she was not looking at him.

Then, as suddenly as it began, it ended. The look on Rena's face transformed from Miss Vulnerable to Ms. CEO, right before Brad's eyes.

He was dumbfounded. He did not know that women sometimes do things like this simply to verify their desirability if their confidence begins to wane. And then again, they also do things like this sometimes because they feel the need to be loved in the most basic of ways.

Rena was unhappy about having achieved neither.

"Time to get back to work," she said to no one in particular as she got up and left the room.

Brad was perplexed. Rena knew he had this thing going with Lissy. He wondered what she was trying to do before he resigned himself to the thought that women sometimes operate in strange and mysterious ways.

Brad was enjoying his second Coors at his usual table at Grubby's when Lissy finally got enough of a break to converse with him.

"You seem all pleased with yourself, Brad," Lissy remarked. "Things must be going well at work."

"I didn't know I was being so transparent," Brad responded. "But yes, things are going rather well indeed."

Then he explained. "We saw a demonstration today of a miniature drone that is sure to fit our needs exactly. It has the size of a rather large moth, the invisibility of a stealth airplane, and the capacity to carry our electrical and surveillance equipment. Best of all … and this is sure to punch the 'buy button' on the folks at the Department of Defense … it can self-destruct if there's a danger of it falling into enemy hands."

"Wow! It sounds complicated," Lissy observed. "Are you sure it will work?"

"Death and taxes, Lissy, but we're on track working out the things that need to be worked out, and everything looks promising."

"Hold that thought," Lissy said as she went away to wait on some impatient customers.

Brad watched her short skirt swishing as she left and was somehow reminded of Rena and her earlier tempting, except this time there was no reason to hold back. He asked Lissy when she came back to his table if she would like to stay at his place that night. He promised a tasty late supper would be waiting for her when she got off work.

"Are you propositioning me?" Lissy asked.

"Do alligators pee in the pond?"

Chapter Four: THE LISTENER

The Listener, covered with sweat, lurched into a sitting position. He cursed the recurring nightmare that kept him from restful sleep. His real brush with death had been stressful enough, but in his dreams, the vision went beyond his last minute effort to save himself, and he actually encountered death and dying in the most bizarre and horrifying ways. Moments ago, he had met death when Minotaurs began piercing his body with their horns and started to eat away at his flesh.

The gruesome endings always caused him to suddenly awaken—thankful for an end to the nightmare.

The beginnings were always the same.

Others were pursuing him in boats. Their boats were slightly faster than his and he was trying to hold them at bay with his revolver. He needed to see where he was going and he could not see forward and shoot behind at the same time. It was only when the boats seemed to abandon the chase that he realized his circular course was taking him directly into the side of the old tanker Lucifer *from which he had just escaped.*

With only seconds to spare, he dove headlong into the water. He did not know how deep he had gone before the shockwave from an explosion racked his body so hard he temporarily lost consciousness. Dazed and disoriented, he found himself at the surface with a piece of floating debris close at hand. It was an old box of some kind that provided support to hold onto, and shelter from view of the boats that now were searching for signs of life.

The latent hydrocarbon fumes inside the old tanker Lucifer *had exploded in a huge ball of flames and the vessel was no more. Only some small, still-burning objects attested to the fact that anything at all had been there before.*

Lucidity returned when a boat was seen heading his way. It was a Boston Whaler with Brad at the helm.

He took a chance that the box he held onto was empty and full of air when he ducked inside. He was in luck. Not only was there space for his head, cracks in the box allowed him to see what was going on around him.

He watched as the Whaler slowly circled his hiding place. He was prepared to duck underneath the water if an attempt had been made to look inside the box. After all, he had just escaped this man's pursuit and he did not want anyone, especially Brad, to realize he had cheated death.

It took seven grueling hours to reach the shore undetected. He would spend the night on the beach so he could straighten his appearance before hitching a ride back to Medlinton where he would find clothing and personal items in his rental locker. He would begin his life anew.

The Listener's nightmares were so vivid; he could hardly separate them from reality. But the reality of his life since the 'accident', as he thought of it, had been anything but a nightmare.

He had decided it would be best if he left his apartment the way it was and never return. It would suit his purposes best if his name remained listed as 'presumed dead'. This would give him the latitude he needed to make the moves he planned to make in order to fulfill his promise to see an end to Brad and his meddling ways.

He soon discovered that although he had tried to kill Brad, there were no 'WANTED' posters out for him, so he was able to move about as he wished although he shunned going to the same places he frequented before.

He still had plenty of money in the bank thanks to some highly profitable experiences in the past. This suited him just fine. He wanted to be able to concentrate on the things he wanted to do without worrying about making a living.

Over a period of several weeks, he formulated a plan to make Brad pay for the suffering he had caused. He would need an accomplice inside Brad's circle of friends and acquaintances, not only to keep him informed of Brad's plans, but also to help him when the time came for the action it would take to eliminate Brad forever.

His accomplice came from an unexpected source.

Once the Listener and his Accomplice had agreed to cooperate with one another, it was not long until Powervert's plan to develop and market an electrically powered surveillance drone aircraft became known. Armed with this information, the Listener could see a way that would give him more satisfaction than murder. He would set a trap for Brad that would expose him as a traitor to his own country. This would be the perfect way to get even and perhaps also send Brad to prison and maybe death by lethal injection.

When the Listener's Accomplice set the microphone under Brad's favorite table at Grubby's Tavern, it was like the icing on the cake. For some reason Brad had a penchant for disclosing the most closely held company secrets to Lissy. Clearly Brad saw that Lissy posed no threat to the company or he would not have said so much in a public place.

Then when Brad thought to have a company meeting at Grubby's, the Listener could not have been more pleased. Now he was privy to more than only the things they were doing. Now he heard how they planned to do them.

With the words from Brad's last conversation with Lissy still ringing in his ears, the Listener called his Middle Eastern Contact.

"Something new," he said without elaborating. "Divide the regular time by two and subtract six hours." He replaced the receiver in its cradle.

The Listener's Contact had understood the instructions perfectly. He had correctly adjusted the regular meeting time of each Thursday at 4:00 PM, the 88th hour of the week, to the 38th hour. The Contact walked into the Medlinton Train Station at exactly 2:00 PM the following Tuesday. They took seats close enough to one another to hear well, but not so close as to be associated.

"Our friends are getting close," the Listener said without making eye contact. "All they need to do is put the package together. Already they have all the parts. The drone is no bigger than the palm of your hand and it can carry nanoscale cameras, GPS, electric motor; everything."

"I hope you can arrange to secure one for us," the Contact said.

"Like I said, it's not completely put together yet. When it is, a way will be found to make it ours. In the meantime, you should inform your organization that their training exercises can continue the way they are for the time being. They will still need to hide their operations from the large-scale drones, but so far that has not been difficult to do. It will be a different story when the small sized drones take to the skies. They cannot be easily seen and they can remain in flight indefinitely because they get their electrical supply from the ground wirelessly."

"We must have one of the models," the Contact reiterated.

"Be patient, my friend, the situation is well under control."

"Double the hour next time plus six?"

"Plus six first, then double," the Listener corrected.

It was quite by accident the discovery was made at all. Brad, sitting at his favorite table in Grubby's, dropped the ballpoint he was using to scribble notes to himself while he waited for Lissy to get a small break from her ever-demanding patrons. He glanced up after retrieving the pen to make sure he avoided cracking his head on the table when he saw something attached to the stem near the top. He was safely re-seated when it occurred to him that the object could be something more than part of the table's construction. Rather than peering under the top to verify whatever was under there, he slipped his iPhone out of his pocket, guessed where it should be aimed, and pressed the camera button. He had guessed correctly. The object was found amongst his photos almost centered in its frame. Brad brought the picture in closer by touching the iPhone screen with his thumb and forefinger together and then spreading them apart. The object was clearly a microphone.

Now Brad was just as pleased that Lissy was still busy because it gave him time to think. His first thought was that microphones were probably under every table. Maybe this was management's way to stay in touch with the customer's needs by listening to their complaints. He rejected this thought when he considered how many people it would take to listen to all the conversations going on, especially on a busy day. He could not contain his curiosity. He made a trip to the men's room and dropped his pen along the way. No other table could be seen to have its own listening device.

By the time Brad returned to his table, Lissy was there with a fresh can of cold Coors. "Hey Lis," he began, "you know this is my favorite table. Is there someone else who favors it also?"

"No, why?" she answered.

"Just checking to make sure no one else is moving in on my territory," Brad smiled.

"No one else has a chance … unless he is richer … or taller … or better look …."

"OK, OK, I get the message," Brad interrupted as he admired Lissy's white teeth sparkling through her smile and the colorful tinsel adding glitter to her hair. He thought he was a lucky guy to have the inside track with this delightful and beautiful woman who also had a sense of humor.

Brad wondered if she knew about the microphone although he did not mention it. He was not even sure whether it might be a toy, or if it was real but did not work or—. If it turned out to be the real thing in working condition, one thing was for sure—Powervert and their industrial secrets had been compromised. He had to find a way to determine whether the microphone was operational, and then to figure out the intentions of whoever put it there.

Brad had no feelings of distrust toward Lissy and he could have simply asked her about the microphone under his favorite table, but something inside kept urging him to keep the knowledge of the discovery to himself. He would have to think of another way to investigate the mystery.

A scanning device could detect the presence of electronic signals and the laboratory at Powervert had a number of these that the scientists used to verify the presence of any electrical current that was being sent wirelessly. As safe as power-beaming seemed to be, Jason and his team would be foolhardy to inadvertently walk into an unseen bolt of electricity in the course of doing their research. It took Brad only a moment to rule out asking Jason for the use of the scanning equipment. Jason, like Lissy, was believed trustworthy, but Brad's notion to play this one solo prevailed.

It was easier than Brad had imagined finding a device that could detect whether a 'bug' was in the area. His visit to the local 'spy shop' store was rewarded with a choice of detection devices from ultra sophisticated to simple. He chose one that resembled a ballpoint pen. It could be easily switched on and off to conserve battery power, and it warned of an electronic signal in the area by flashing an LED in the barrel of the pen.

He could hardly wait to use it at Grubby's, but he decided to try it first in the laboratory at Powervert. Electronic signals were sure to be found flying all over the place at the lab and besides, if it failed to blink at Grubby's, it could either mean that the microphone was dead or that the pen did not work.

It worked all right.

Even before Brad went through the door to the lab, the LED began blinking away industriously.

Next stop; Grubby's.

Brad planned his investigation for midweek when business would be at its slowest, and for a time when Lissy would be near to closing for the evening. Lissy saw him walk in and met him at his favorite table with a can of Coors and a frosty glass.

"You're late! I was wondering if you were going to show up."

"I wouldn't miss my Lissy fix for anything. Besides, I worked late at the office," Brad lied.

"You're sweet, but you're going to be limited tonight. I plan on closing soon."

"This one to get the dust out of my throat and one more to relax with and I'll be just fine."

"That'll work," Lissy said as she left to begin her closing routine.

Brad knew that the process of closing up shop would take about thirty minutes and that he and Lissy would leave together, grab some tacos or a bucket of chicken, and head for his place.

He reached into his pocket and pushed the button of the scanning pen. The LED began blinking immediately. Brad got up from the table and began moving away. After only a few paces, the light stopped blinking and Lissy called out, "What are you doing?"

"Just stretching my legs," Brad answered. "I've been sitting all day."

He began to wonder if these lies that came so easily were a negative reflection on his character. But he had accomplished the job of justifying his movements and he continued wandering around the room searching for the receiver he was sure to be nearby—recording the words the microphone collected.

And there it was!

As he passed by the door where the bar's liquor stock was stored, the telltale blink began anew. A cloud of doubt passed over Brad's mind.

This room was always locked and the only people with access were Lissy and the other bartender who was only called in from time to time. Brad supposed that the owner of Grubby's also had a key, but the store of spirits would surely be secured from all the other employees, especially the cooks who had a reputation for using an inordinate amount of liquor in the preparation of food. There was never any doubt as to who was getting the 'preparation'.

Now Brad was worried. Lissy had no reason to spy on him. He always kept her alerted to the things Powervert was doing—unless—unless Lissy was using the device in the storage room to record information and then taking it or sending it somewhere else.

Suddenly, Brad wanted to be alone. It would not be possible for him to think clearly about Lissy and the microphone mystery with her nearby. It had always been difficult for him to think of anything else when Lissy's perfume filled the air and the warmth and softness of her body was within reach.

A look of bewilderment came across Lissy's face as Brad announced, "Sorry Lis, I'm going home. I don't feel well."

"I'm sorry to hear that."

"It must have been something I ate. I'll stop by tomorrow." And with that, Brad walked out the door leaving his beer, his unpaid bill, and his flabbergasted friend behind.

His lies were becoming so frequent and so good he was thinking of turning pro.

Brad did not go back to Grubby's Tavern after work the next day or the following day. When Lissy called to make sure he was OK, he told her that he was still feeling poorly, but the doctor had told him he would live. One lie after another; when would it all end?

Just then it occurred to him that his conclusions about Lissy's involvement could have been made too hastily. He had judged her guilty—at least at some level—based on circumstantial evidence and circumstantial evidence is often misleading.

The electronic device locked up inside the liquor storage room could be recording information and it could also be receiving and transmitting the information to another location. Brad wondered why he had not thought of this before. He was hoping with all his might that a transmitter would be discovered to help him re-connect with Lissy. She would not be completely out of the woods suspicion-wise, but at least she would have a way out.

Brad made his way toward Grubby's at his usual time. Except now, he walked around to the alley behind the building. The sun was going down and the light in the alleyway was dim, but Brad knew about where he would find the outside wall of the inside storage room. It was clearly identifiable. As Brad approached, he was able to distinguish a small window set ten or twelve feet high in the wall. The window was in the open position, but steel bars set vertically in the frame would keep even the smallest person from entering.

Brad retrieved his scanning pen and activated the switch. It began to blink. He was aggravated with himself for having achieved nothing. He still had no way of knowing whether the device was recording or transmitting information. The window was too high to see into and besides; it was almost dark outside and even darker inside the storeroom.

Brad decided to try to take a look anyway. A small dumpster nearby would provide a platform and its wheels allowed it to be moved into place easily. He had a little difficulty climbing aboard the dumpster, but once he stood atop it, he was at eye level with the window. With his eyes becoming increasingly accustomed to the darkness, Brad peered into the room. Yes, it was the liquor storage room. Brad could see the dim outline of bottles on each of the shelves. They were some distance below the window and well beyond the reach of any would-be thief. But, higher up and not far away, was another row of shelving. On this row and about an arm's length away, a small box-shaped object seemed to emerge from the gloom. Located where it was, it would not be possible for someone inside the storeroom to see the box unless they stood on a ladder. But it was possible for someone, like Brad, outside to reach in and retrieve it. Brad reached in.

Suddenly the dumpster Brad was standing on began to move in a way that caused Brad to lose his balance. It seemed his leaning one way caused the dumpster to scoot the other way and Brad fell to the ground—hard.

He would not be sure if the vision he saw was a part of the happening or a nightmare about it. He thought he saw a dark figure looming over him as he lay in the alleyway. Then his consciousness fell into a black abyss.

Everything seemed fuzzy. Brad had no idea where he was. Clarity returned only when Brad heard the stranger ask if he was all right.

"I heard your cell phone ringing or I wouldn't have known you were here," the man said.

"I'm glad you did," Brad said sincerely.

"I was on my way into Grubby's," the stranger offered. "Can I help you inside?"

"Oh no, I'm quite all right, but I'll go in with you. I hope you won't tell anyone you found me outside in the gutter," Brad pleaded.

"Don't worry. I've 'slipped and fallen' before myself," the stranger joked as he held the door for Brad then headed for his friends waiting for him at the bar.

Lissy rushed over to Brad when she saw him take a seat at the end of the bar. Things just were not working out well for Brad. First it was the incident in the alley and now some strangers had occupied his favorite table. Oh well, they would probably be leaving soon. He made a quick check of his appearance as he saw Lissy heading his way.

"I've been trying to call you. I've been worried," Lissy began.

"I knew it was you calling. I was just outside talking and I didn't want you to waste your cell phone minutes when I was coming right in." Was he becoming addicted to lying? Maybe he would never tell the truth again. The lies had been told for a reason when he was not sure about Lissy's involvement in the eavesdropping thing. But now he had seen the recorder or transmitter or whatever it was in a place where she could not have seen it or reached it, his suspicions were dissolving and he had no reason to express his distrust through lying.

Brad did not understand his reasons for excluding Lissy from his inner circle, but he still regarded her proximity to the microphone under his favorite table and the box in the storeroom sending or receiving signals as a basis for suspicion. And then his suspicions were lifted.

When he got up from the bar to make a trip to the men's room, he thought to activate the scanning pen in his pocket. The flashing signal from the storeroom was gone. Someone had turned the device off or removed it, and it could not have been Lissy.

"I just talked to Hector on the phone," Brad told Lissy the next time she came near where he was sitting. "He said he hasn't seen you in a long time and that he would fix you a late supper if you wanted to stop by after work tonight."

Lissy's laugh was a promise that Hector would get his wish. Then she verified her attendance by saying, "I don't know how anyone could refuse and invitation from such a world renowned chef who enjoys eating as much as his guests do."

Chapter Five: MORE NIGHTMARES

Once again the recurring nightmare jolted the Listener awake. This time he was thankful for the interruption to his sleep because the bad dream had only begun and he was spared the agony of the entire macabre scenario. The feeling of relief was overshadowed however by his recollection of Brad and the way he had become the Listener's nemesis.

The Listener had assembled a group of buyers in a barn situated in the countryside a few miles from Medlinton. The buyers were Middle Easterners who were observing a laser-based weapon—over which the Listener had control—that would fire a lightning bolt of electricity. Whether the buyers were interested in the weapon because they intended to place it in the hands of terrorists was of no concern to the Listener.

Somehow Brad had learned of the demonstration about to take place, and had attempted to foil their plans by calling the police and setting fire to the barn after bolting all the doors shut. With no way to get out of the burning barn, all the others in attendance were burned to death.

The Listener was the only survivor. He had barely been able to save himself by jumping out of the loft door at the rear of the barn—breaking his leg as he did so. Fortunately, no one had seen him jump and he was able to escape across the valley under the cover of darkness.

Since the police report had stated that all those who were inside the barn died in the fire, the Listener was free from pursuit, and he could move about as he wished. And every move he planned from that time forward was calculated to take him to a position where he could exact his revenge on Brad for trying to burn him alive!

The Listener's opportunity for revenge came when he found out that Brad and some others were due to visit the old tanker *Lucifer*. The Listener set up his power-beaming weapon in an attempt to kill Brad with a lightning bolt of electricity. The attempt went horribly wrong, and the Listener was forced to escape the scene on his own boat.

That was when he lost the lightning bolt weapon as his boat circled back and crashed into the tanker. The nightmare was back. There was no escape. Brad was responsible and he had to die.

The Listener was fearful of going back to sleep. It was only 4:00 o'clock in the morning, but that did not matter. He was not going to put himself in a position to begin the nightmare anew. He got out of bed and turned on the shower—cold water only.

In spite of the hour, the Listener was refreshed. He dressed for the day, primed the coffee maker, and went to his study to listen to the latest recording. He was not prepared to hear the voice of his accomplice, but there it was:

I had to remove the recorder ... I saw Brad on top of a dumpster reaching for it. I didn't know what to do. I pushed the dumpster hard and Brad fell to the ground. I made sure he stayed there by kicking him in the back of the neck. He may have gotten a glimpse of me, but there was no way he could have seen me clearly in the dark ... I'll put the recorder in a plastic bag and leave it behind the shrubs by your front door ... I, um, don't think it would be a good idea to continue with our arrangement. You already have all the information I know about, and if Brad can get close to finding out about us, others can too. Please don't contact me anymore.

The headset went silent.

Later that afternoon, the Accomplice found the following message on his telephone voicemail:

Your request for disassociation is understood but cannot be granted. The journey taken so far has brought to light all the peripherals, but not the product itself. Please see to it that a full sized working model is delivered to me before the next ten days have passed. Please do not burden me with any difficulties you may encounter fulfilling this stipulation.

The Accomplice had no difficulty understanding the implied threat.

At first, the deal he had made with the Listener seemed like a win-win situation. He would provide recorded conversations between Brad and his associates in exchange for the promise of untold wealth in the form of a high paying position with a new company that would use the technology gleaned from Powervert to generate an unlimited amount of money from the licensing and sale of the products. And since the Accomplice had helped to create the product, it was only fair that he should be able to reap a greater share of the success than Powervert was willing to give him. Planting the electronics to make eavesdropping function was easy. Getting the information was even easier because Brad had a tendency to talk openly to Lissy and others who would gather at his favorite table.

Lately, the Accomplice had been having second thoughts. Although it had not been discussed openly, he recognized the likelihood that the technology would be sold to terrorists once it was gathered. If the terrorists planned to use the information to thwart others spying on their operations, he could live with that, but it was repugnant to think that the terrorists would use the technology to aid their random murder of innocent bystanders.

Now that the Accomplice believed the Listener had all the information the spying activity was likely to produce, he wanted out. He did not care about the promises. He thought it would not be possible to live with himself if he followed through with his plans. He became fearful that deeper involvement might lead to the point where it would not be possible to quit without becoming the victim of some sort of reprisal.

Maybe the point had already been reached. The voicemail from the Listener had definitely left the impression that the Accomplice would not enjoy the things that would happen to him if he failed to provide the working model as ordered. He set out to deliver the MothBot to the Listener before the deadline of ten days passed.

Derek was as obliging as he could be. He understood perfectly well why several complete models were needed. Yes, it made sense to be able to work on different aspects of development simultaneously. When the models were delivered to Powervert, the Accomplice would make sure that two of them would somehow be destroyed with their self-destruct capabilities. Of course only one would meet its fate this way. The other would be taken from sight and modified with all the features needed to call it a working model.

Chapter Six: SPY TIME

The late supper that Hector had promised materialized in the form of a pepperoni pizza from Pizza Hut and a bottle of Santa Margarita Pinot Grigio from the local liquor store. Lissy did not care. She thanked Hector profusely for his culinary knowledge and abilities. Later, she thanked Brad by providing one of the best bedroom experiences he ever had. The few days they had been apart must have magnified the intensity of the occurrence and Brad was satisfied completely when the time came to sleep.

The peace that comes with sleep, however, was not to come to Brad this soon. He laid awake for a seemingly endless amount of time thinking about the events that had befallen him. Whoever had gotten the classified information from him was in a position to undermine Powervert's efforts at developing the MothBot. Brad was not sure how the proprietary information would be used, but he was reasonably certain that it would not be beneficial to Powervert.

Brad decided his only course of action would be to discover the person guilty of this industrial espionage and to put a stop to whatever the person was up to. Now Brad began to wonder if Hector could read his mind. Because, even though Brad was not conscious of Hector's presence, he suddenly felt a gentle kneading on his shoulder and he could hear the soft sound of purring.

Brad's investigation began mentally while he lay awake as Lissy slept soundly beside him. He thought it had to be someone close to him whoever it was. He thought of his closest associates. Hector purred and kneaded.

Brad thought that Lissy was the most likely because of the circumstances, but she was probably not the one. Hector continued purring and kneading.

He thought the guilty party could be one of the other company officers, Rena or Jason. Hector became motionless and silent.

He thought of the members of the research team. Hector purred and kneaded once again.

So, Brad thought, *according to Hector it is one the scientists. Each of them would certainly be in a position to gather all the information directly from their activities with the company. What reason would they have to plant a bug?* Hector was no help. He had left for the kitchen to get a midnight snack.

When sleep finally came, it was fitful and constantly interrupted with nightmarish, obscured visions of ghostly figures and flashing scenes of fire and smoke and devilish faces. The faces were easily recognized as those of his friends and associates, although their features were grotesque and distorted with sinister smiles or grimaces. Some of the faces seemed to be laughing at him while others were forlorn and sad. As the fire and the smoke and the faces moved randomly across his dreaming field of vision, one shadow of a person seemed to pause over him as though it was the Devil himself about to descend.

Brad awoke with the specter of this ethereal vision still fresh in his mind. The cobwebs of sleep were soon brushed away as Brad fully awakened and reality returned. There would be no more sleep this night. The graying dawn provided enough light for him to slip out of bed without disturbing Lissy, and avoid tripping over Hector who was anxious to join anyone headed for the kitchen.

Brad's first action was to get the coffee pot in preparation of a cup of the hot stimulant, but his actions were interrupted by Hector weaving back and forth between his legs in anticipation of a morning snack. Hector won out, but the coffee pot was soon bubbling away and the aroma of the fresh brew filled the air.

Remembering the vision in his dream, Brad searched his memory for a face to fit the body. None was forthcoming. There was only the silhouette hovering above him. But there was something. Yes, the figure seemed to be dressed in a suit. It could not have been the man who helped Brad to his feet beside the dumpster. That person was wearing a polo shirt with no jacket. Brad wondered who he knew that wore a suit. The customary attire at Powervert was much less formal than to require business dress, so it was unlikely the shadow that had hovered over Brad was a close associate. But—yes—Jason did sometimes wear a suit and he was the only person Brad could think of who did so on a regular basis.

Although Hector had not kneaded his approval when Brad considered Rena and Jason as suspects, no person close to Brad other than Jason fit the vision of the shadowy figure. Hector could have been wrong, but Jason, as one of the owners of Powervert, did have direct access to all the company's inside information.

Although Brad considered the odds to be negligible of the finger-of-guilt pointing toward Jason, he decided to monitor his activities anyway, just to see if anything would turn up.

Only two days later, Jason told Brad he would be working at David's Place for the rest of the day. David's Place was not David's place at all, but that name had stuck after the incident at the barn. David Drendorf was working for Powervert at the time and he lived in the farmhouse next to the barn. The barn that was 4.3 kilometers away and across a valley from Powervert was used to test reception of their power-beaming experiments, and it had burned to the ground in a tragedy that had cost David his life.

Brad decided to take advantage of the absence of Jason to look around his office even though he felt a tinge of guilt snooping this way. As expected, he found it neat and tidy.

Of course it should be because Jason had the opportunity to get as messy as he liked in his laboratory. Brad moved around behind the desk. The file drawer was the only one he found to be locked. The contents of the other drawers were as organized as the space on top of the desk, and nothing of a suspicious nature could be seen. The contents of the credenza against the wall behind the desk were examined next.

When Brad opened one of the lower doors, the first thing he saw was a recorder! He was not exactly sure it was the same one he had seen in Grubby's liquor closet, but it certainly could have been. Brad's suspicions were further aroused when he realized the device was not just a VHS or DVD home recorder. This one was a professional model made by Scientific Atlanta. He was curious to know if the device was capable of receiving and transmitting remote signals when he heard the sound of a door opening and closing. Rena must be coming back from lunch early. Brad closed the credenza door and returned to his own office.

Although Brad rarely closed the door to his office, he did so this time. He needed time to think without distractions. The discovery of the recorder in Jason's office had convinced him that he was on the right track leading to the person who would sell-out Powervert.

Brad decided to approach the situation logically. He would ask himself to answer the 'Five W' questions: Who, What, When, Where, and Why. He would not have to rearrange the order to suit the things he knew and those things unknown. The first four 'Ws' appeared to be known. Jason (Who) had acquired proprietary industrial knowledge (What) in the recent past (When). The information could have been taken from Powervert itself or through the listening devices at Grubby's or both (Where). The only remaining 'W' was Why?

A reasonable assumption would be to suggest that money—under the impetus of greed or revenge—could provide enough incentive for someone to become a traitor. This thought brought back the 'Who' part of the original query. Who would be the recipient of the industrial knowledge? Brad realized a second set of 'W' answers would be required in order to get closer to solving the mystery. The person—whoever the second 'Who' was—must have paid Jason for the proprietary knowledge in the recent past. But the second 'Why' was also unanswered. Jason was privy to all the information Powervert had. Why would he plant a listening device? Why would someone compromise Jason's position with Powervert for the information? The finger of guilt pointed more assuredly at Jason as Brad realized the microphone and the recorder could have been planted at Grubby's for no other reason than to throw others off the culprit's trail.

Now that Brad was confident he had solved the industrial spying mystery, he decided to see if he could use Jason to lead him to whoever it was that was receiving the information. There were indeed many of Powervert's competitors out there that would place a high value on this kind of industrial knowledge.

Brad knew the business advantage of being the first responder with any technological innovation. The first company to introduce a product did not always win, but it definitely had a great advantage over its competitors because they would be forced to play catch-up.

Brad decided to follow through with his plan by doing the same thing Jason did—eavesdrop. He revisited the spy shop that had provided his electronic signal-detecting pen. The people there suggested and sold him a recorder that could be connected to the company telephone system to record all the conversations that came over the lines. The device was soon installed in Brad's office and locked inside his credenza. Then the waiting game began.

The ringing telephone startled Brad with the thought that some special information could be incoming before he realized that it was his phone that was ringing. He called himself an 'idiot' as he picked up the receiver. It was Rena calling from the front office.

"Brad, a delivery person is here with a box for Jason. Do you know where he is?"

"No," Brad answered. "Why can't you receive it for him?"

"Well, the guy says he was told to deliver it to Jason and that's what he wants to do."

"I'll be right out," Brad said in resignation.

It turned out the delivery was from Dronomics and after much discussion with the delivery person, Brad was able to sign for and receive the box.

"See if you can get Jason on his cell phone and ask him what he wants done with this," he instructed Rena before returning to his desk and continuing with his day's work.

Only a short time later, Rena told Brad that Jason had been contacted and the box was to go to the laboratory where the crew there would know what to do with the MothBots it contained.

"Did you say MothBots like in plural?" Brad asked.

"Yes, I understand there are five of them in there," Rena responded.

"OK, thanks," Brad said as he wondered why Jason needed so many drones.

When he got the chance to question Jason about it, he replied that the scientific team was attempting a number of experiments at the same time and the five drones would allow the process to be expedited.

The ability of Slade, Gordon, and Bratley to perform their experiments more efficiently soon came abruptly to an end. Two of the MothBots had somehow self-destructed the way they were designed to do, but this time it was accidental.

Brad was not happy about this. What sort of dependability could be attributed to equipment that could self-destruct so easily? Jason could not provide answers. It seemed that no one was present in the laboratory when it happened. All that was found the next day were some burnt and distorted parts and hardware. However, Jason assured Brad that the remaining MothBots could be used to satisfy the research necessities until the missing ones were replaced, and that special precautions would be taken to see to it that no more accidents of a similar nature would take place.

The Accomplice held the treasure in his hand. The MothBot was a perfect specimen of technology and innovation. He carefully opened the fuselage to expose the elements inside. A motherboard containing a GPS receiver was used to control all the MothBot's functions. Power to the rotors came from a small electric motor no larger than a dime in circumference. The electric motor got its power from two Lithium Hydride wafer batteries and a miniature rectenna designed to receive electrical power from a remote source and redirect it to the motor. A multi-faceted servo operated the flight controls. All the components were hermetically sealed to prevent an incidental spark from the operations to ignite the hydrogen that would be injected into the body of the craft after it was closed up and ready to fly. The hydrogen could be ignited with another component designed specifically for the purpose. When activated, a piezo striker would cause the hydrogen to ignite and the MothBot would self-destruct.

Anticipating the need for a controller to operate the MothBot, the Accomplice had duplicated the Dronomics original in his spare time with no one else's knowledge. The package was now complete. He would arrange to deliver it to the Listener and all his troubles would be over. The payments that had been promised for his cooperation had already begun.

He and the Listener had agreed on five hundred thousand dollars to be paid at the rate of twenty thousand dollars per month for twenty-five months. The Accomplice was certain the payments would continue after the drone and the operating device were delivered. After all, the Accomplice's portion of the agreement would have been completed and the Listener would be bound to honor his obligation.

In spite of his self-assurances, the Accomplice was just a little worried. He had placed himself in a position of vulnerability when he agreed to help the Listener. It seemed innocent enough when the two first met ....

The person sitting next to him at the bar one afternoon surprised him by identifying him by name. The person introduced himself as a special investigator for the company that provided insurance for Powervert. He said the incident at the barn had caused his company to lose a great deal of money, and the fire had been started under suspicious circumstances.

The man refused to tell the Accomplice his name. He said that he was known by too many people at Powervert to do that or to personally participate in the investigation. He did, however, convince the Accomplice that he knew what he was talking about by providing information that no one outside Powervert's inner circle would know.

He said Brad was suspected of criminal involvement. All he wanted was to hear him talking in an atmosphere that was not business oriented, so he might say things that would otherwise not be said. He said if the Accomplice would place a microphone at Brad's favorite table and a recorder in the liquor closet at Grubby's Tavern, his company would pay him five hundred dollars per day.

Since Brad visited Grubby's almost every day, the surveillance only needed to last for about two weeks. All the Accomplice had to do during that time was to collect and replace the tape after each of Brad's visits. If the surveillance needed more time, the Accomplice would receive daily payments accordingly.

What could be easier? the Accomplice thought as he agreed to the deal.

That was only the beginning. Once the Listener had discovered the magnitude of the project currently underway at Powervert, the criteria changed. The value of the inside information about the research went from seven thousand dollars to five hundred thousand dollars. The new money was to be paid at the rate of twenty thousand dollars each month beginning immediately and continuing until the entire amount was paid.

The Accomplice became anxious to get the money in spite of his misgivings about the amount being more than the value of the information unless it was to be used in a harmful way to Powervert or some other less-than-legal or unethical way.

When he questioned the Listener about the use of the information, he was told that the amount of money he would receive was miniscule compared with the losses the insurance company had taken and they were hoping to be able to recoup some of it.

Something about the way it was stated made the Accomplice doubt the veracity of the things the Listener told him. Surely a barn—even if it was loaded with technological equipment—would not be worth that amount, but he decided he was already in too deep to back out now. Besides, he believed that Powervert had deprived him of his fair share of the credit for developing the MothBot. He told himself that he was not out for revenge, he only wanted what was coming to him, and so he went along with the Listener's plan.

… That was the way the whole thing began. It was simple at first and then became more complex as time went by and circumstances changed. Now it had gone too far. The accomplice was convinced that the information he was providing the Listener was going to be used for sinister purposes.

He had no solid proof, but he knew the capabilities of the MothBot as a remote vehicle for the surveillance of terrorist activities, and he was aware that the terrorists themselves could use the technology to counteract the MothBot's purposes.

The Accomplice's suspicions were justified when he learned the Listener might not be who he said he was.

He happened to walk into the Medlinton Train Station one afternoon just after 4:00 PM when he saw the Listener sitting on one of the benches. The Listener did not see him, so he took a position where he could observe what was going on without being seen. The Listener was clearly communicating with the person sitting next to him, although they were acting as though they were strangers. Something underhanded was sure to be happening. The Accomplice had no way of hearing their conversation, but it was obvious that both men were of Middle Eastern descent.

As much as the Accomplice frowned on stereotyping, he could not help from linking much of the world's terrorism activities with this group. And the fact that he was supplying the Listener with knowledge of anti-terrorist technology made the linking more realistic.

The subject of murder had never occupied much of the Accomplice's time. Naturally, the subject came up from time to time, for example when people were arguing about the justification of the death penalty. He did not think that murder was justified except in cases of self-defense, but he also believed that a person who was guilty of murdering someone else should not be allowed to go unpunished.

The Accomplice was aware that there were valid arguments for and against capital punishment on both sides of the issue, although he would not have felt strongly enough about it either way to participate in a debate. However, there was one aspect of the taking of life about which he felt defiantly adamant—the murder of innocent people.

In the Accomplice's eyes, Nature already had made sufficient arrangements for innocent people to lose their lives by providing natural disasters. Each year thousands would die from earthquakes, avalanches of mud or snow, fire, tsunamis, disease, and other forms of indiscriminate 'cleansing'. Interference with Nature's Plans by ordinary human beings was thought of as intolerable. The Accomplice could imagine no circumstance where random death was justified and here he was, possibly contributing to the very thing he despised the most.

When these realizations became apparent, the Accomplice resolved to distance himself from the Listener and his ilk. The fact that he could possibly be wrong in his assumptions was of no consequence. He did not want to even come close to participating in any form of terrorist activities.

That was when he went to the alley behind Grubby's to remove the recorder and break his ties with the Listener. His surprise at finding Brad snooping around only strengthened his decision.

Chapter Seven: SECOND THOUGHTS

With the MothBot and controller in hand, and the ten-day's notice about to expire, the Accomplice dialed the Listener's cell phone and left a message:

The items you requested will be delivered tonight
behind the shrubs as before.

Brad intercepted the call. It had originated in Powervert's laboratory and Brad was perplexed by his inability to identify the caller. Obviously, a handkerchief or some other object had been placed over the telephone mouthpiece to disguise the voice.

This could be the break Brad had been waiting for. He would follow Jason with the hope that he would be led to the person who was paying for Powervert's proprietary information. There was only one problem. The call had been made from Jason's laboratory and Jason was working at David's Place—or was he?

Brad called Jason on his cell phone under the pretext of needing some information related to their business. Jason was indeed at the remote site everyone called

David's Place. Was it possible that Jason was forwarding his calls somehow from his cell phone to the phone in the laboratory, or could it be that one of the other scientists, Slade, Gordon, or Bratley, had made the call to the mystery person?

Brad could have found a way to discreetly follow Jason's activities, but it was not possible for him to follow all four members of the scientific team—unless—unless he did not follow them at all, but simply verified their evening activities.

The telephone message had noted that the items, whatever they were, would be delivered tonight. Brad would monitor the homes of each of the scientists to see who was home during the evening and who was not.

Brad's plan called for driving a circuitous route passing by each of the four homes he was monitoring in a continuing pattern. He would alter the route from time to time to avoid arousing suspicion from anyone observing him, and he would occasionally stop for coffee to break up the timing of his visits to each of the locations.

The first home Brad passed, Gordon's, provided few clues because his car must have been in his garage, but the house lights indicated a level of activity that probably meant Gordon was at home.

Checking on the other three scientists was easier because Jason's, Bratley's, and Slade's cars were parked outside near their residences.

The second and third times around, nothing appeared to have changed. The fourth time Brad made the circuit, a little after 9:30, Jason's car was missing. The car reappeared again just before the midnight hour—the time Brad had decided he would end his vigil. He had not seen Jason's car leave or arrive, so he could not know whether he traveled alone, or even if it was Jason in the car.

So, Jason must have forwarded the phone call from his cell phone to the lab phone to throw any possible listeners off the track, and to eliminate the possibility of the number he called showing up on his call record, Brad thought.

Brad was pleased for having identified the information-stealing culprit, but he was disappointed about not being able to follow him to the person who was buying the technology.

With the question of 'Why' still nagging, Brad decided to take advantage of the late hour to make a thorough search of Jason's office in case there might be some incriminating evidence to be found.

Brad drove to the Powervert building, identified himself to the security guard hired by the industrial complex, and soon was inside Jason's office.

It did not take long to pick the lock on Jason's file cabinet drawer. The contents of the drawer revealed no secrets, although the files inside contained enough information to compromise all the plans Powervert had made.

Once again, the 'Why' question came up. Jason would only need to copy all these files and sell them to anyone willing to pay. *Why would he want to record the conversations?*

The contents of the rest of the desk were as non-telling as the file drawer. It was only when Brad revisited the credenza the mystery deepened. The recorder he had seen there before was still there. The items Jason delivered obviously did not include this device.

The Listener retrieved the black plastic bag from behind the shrubs near his front door. A smile crossed his face as he checked the contents. It was all there. The controller, with its video screen and all its keys and switches, was more than the Listener had expected. And then there was the MothBot that shrouded its structural toughness with fragile-looking gossamer coverings. The Listener was almost afraid to touch it, but when he did, he discovered it to be much more solid feeling than its appearance would indicate, although its rotors were as flexible as one would expect a real moth's to be.

And he recognized two things he had not thought of before: He would have to remain in contact with the Accomplice to answer the number of technical questions that were sure to arise, and the MothBot would present too much of a challenge for him to duplicate without help from someone with the highest level of expertise.

The Listener decided his Contact would have to locate the means to reproduce the MothBot, and whomever he found would have to pay dearly for the privilege. The next meeting with the Contact was not that far away and the Listener would use the time until then to photograph all the MothBot's parts and to learn as much as he could about it. Maybe he could even make it fly!

When the time came for the weekly meeting with the Contact, the Listener went through his usual precautionary procedure of taking a different route to the train station and entering through a different door.

On this day, the Listener was surprised to find that his routine to avoid others observing his actions, allowed him to be the observer. As he entered the station through one of the side doors, he saw a familiar figure standing near the newsstand and pretending to browse the magazines. Even though the man's back was turned to him, the Listener recognized the Accomplice immediately. Quickly retracing his steps to avoid being seen, the Listener exited through the door he had just entered, went around the building, and re-entered the station through the main doors. A glance around the waiting room told him that the Contact had not arrived, so he took a seat away from other people, but where he could see the newsstand surreptitiously. He observed the Accomplice reposition himself so he would be out of sight.

The thought that this could be simply a coincidence crossed the Listener's mind, but after the Contact had arrived and their discussion had been going on for some time, the Accomplice's continued presence, somewhat out of sight, verified that he was spying.

For someone who expressed his wish to become disassociated, the Accomplice is acting foolishly and dangerously, the Listener thought.

The meeting with the Contact was fruitful. After he had been shown the MothBot, he told the Listener that his Middle Eastern contacts had access to government sponsored technological resources of the highest caliber. He expressed certainty that the MothBot could be duplicated in days and not the months the Listener had surmised. The Contact assured the Listener of another aspect of the relationship between them. The Listener would be rewarded well for his participation, but the amount would be determined by the Contact—not the Listener.

The Listener realized he was being given an 'offer he could not refuse' and turned the MothBot over to the Contact. In exchange, the Contact agreed to return the MothBot after it had been cloned along with as many others as the Listener wanted. The Contact stipulated the drones would be for the Listener's own personal use, and no one else was to be given access to the technology.

The conspirators parted when the Contact told the Listener his compensation for the MothBot was sure to exceed seven figures.

The Accomplice was still lurking behind the newsstand.

Although the Listener was privately dissatisfied with the deal that took his prize for an undisclosed amount of money, he reconciled the situation with the knowledge that he still had sole possession of the controller, and that he would soon have the means with which to put an end to Brad Ganderson.

The Accomplice was now convinced beyond a doubt that he had sold his country out by providing the MothBot technology to Middle Easterners who were undoubtedly terrorists. He was beside himself wondering how to take the next step. He realized the greed that he thought would make him a winner had put him in a loser/loser situation.

To be sure, he still had the money coming in from his deal with the Listener, but he would gladly give it all back to be able to free himself from the thought of aiding terrorism. Only a big-time loser would help terrorists.

He could go to Brad and confess his guilt, but what good would that do? He had no proof that the Listener had done anything wrong, and confessing to selling him Powervert's proprietary technology would only cause the Accomplice to lose his job and to lose his reputation— Loser! —Loser!

If only he could think of a way to make things right.

In the meantime, Brad was not waiting for someone to come running to him with a confession. He was hot on the trail of the person who had been spying on him to gain knowledge of Powervert's technology. His suspicions had pointed to Jason time after time. Now he needed some way to catch him in the act, although that might be a little difficult since the recording equipment had been removed from Grubby's liquor closet. But then, Jason did not need recording devices anyway to get inside information from Powervert.

Brad never did understand why he went to the trouble of recording things in the first place. Nevertheless, that was the way it was and little could be done about it now. Brad decided his best chance of catching Jason in the act would be to continue monitoring his telephone conversations with the hope that he would unknowingly provide Brad with information that would lead to the identity of the person Jason was giving or selling the technology to. And, in spite of Brad's wishes to keep his investigation to himself, he decided to take a chance by asking Rena to help him.

He was somewhat reluctant to meet with Rena alone. The last time they had been alone together, had proven to be awkward. But he trusted Rena and respected her judgment of people and her assessment of circumstances. The safest place Brad could think to meet was on the roof of Powervert's building. The two of them had met there before for business discussions of a serious nature and the outdoor setting should remove them from the intimate feelings that might be felt in a private office.

The time had been selected when Jason was scheduled to be working at David's Place. Rena seemed nervous about the rooftop meeting, and this made Brad wonder if he had made the right decision to include her in his quest for answers, but he went ahead anyway.

"It may be none of my business, but it seems Jason and you have been seeing each other a lot. Is there something serious between you?" Brad began.

"You're right. It is none of your business," was Rena's terse response, "but I don't mind telling you that Jason and I are business associates the same way you and I are. We do socialize occasionally, but nothing more. Why do you ask?"

"I'll get to that later, but first I want to say that you and I have been through a lot … trying to hold Powervert together and to make something of it. I don't think that your commitment and dedication has waned, but I want to hear you say that the goals that you and I have for our business' future have not changed."

Something about the sincerity in Brad's voice and the look on his face touched Rena in a way that convinced her that his intentions were not going to be harmful.

She displayed a concerned look as she answered, "The last few years of my life have been dedicated to my work; sometimes at the expense of my social life. Powervert has taken on a fatherly role in my eyes and I feel like I am a part of a family here. I would never allow anything to harm this relationship."

"I knew that," Brad acknowledged. "I only wanted you to verify it before telling you that our company has been compromised."

Brad watched Rena's facial expressions carefully as he delivered this news. He was glad to see that the message came as a surprise to her.

He anticipated her question when he said, "What I mean by 'compromised' is our plans for the future of Powervert and its development of the MothBot have been stolen from us."

"That's pretty serious, Brad. Are you sure?"

Brad told Rena about the way his suspicions had been aroused when he discovered the microphone under his favorite table at Grubby's, and how he had investigated and found the recorder in the liquor closet. He confessed to having suspected Lissy before he realized the recorder could have been placed in the closet without her knowledge, and the person who attacked him in the alley was certainly not a woman. He told Rena how he had hoped to find the person gathering the information so the person who was paying for it could be found as well. The trail had gone cold however; when whoever the guilty party was, discovered Brad was on to them.

"So let me see if I understand this," Rena said. "You say the microphone under the table at Grubby's was used to gather information about Powervert's plans. Would the person making the recordings be able to gather enough information this way?"

"You tell me," Brad contested. "You and Jason were sitting at the table when these very plans were being finalized."

"That's true, but would this provide enough detail to cause one to think that all our plans have been compromised?" Rena challenged.

"There's more," Brad admitted.

"You think it's someone in our company doing the compromising, don't you?"

The look on Brad's face answered her question.

"You think it's Jason, don't you?" A long pause followed this question; again providing Rena with the answer. "Don't worry, Brad. I'm as much for finding the guilty party as you are … no matter who it is."

A look of relief washed over Brad's face. Now he was able to continue with the evidence leading to his conclusions about Jason. He followed through by telling Rena about the wiretap, his after-hours surveillance, and all the other things that caused him to think the way he did. Rena agreed that his conclusions seemed reasonable, but she failed to understand why Jason would need to record the things said at Grubby's. Brad could not explain why except to tell Rena that it possibly could be simply to lead others to conclude it was someone other than a Powervert insider.

"So now I understand," Rena ventured. "You suspect Jason has sold out our company and the evidence leading to him dried up with the loss of the recorder, and now you want me to take over as a spy to gather the evidence needed to verify his guilt."

"That's close," Brad admitted. "What is really needed is evidence identifying the person who Jason is selling the information to. Buying or otherwise acquiring industrial secrets is a crime and I'd like to see this person busted."

"I'll do what I can," Rena said. "Why don't you try to think of a way we can find out the things we need to know while I go downstairs and get a soda? Do you want one?"

Brad agreed it was time to take a break and he tried to devise a plan, but none was forthcoming. When Rena returned with the drinks some time later, Brad told her the only thing he could think for her to do was to be vigilant and hope that Jason would make a mistake.

It was time to relax. The intensity of their conversation had exhausted them both, and they sat in their respective chairs sipping their sodas in silence. Brad was trying to think of something to say, but no words came to mind. He glanced toward Rena to find that she was looking directly at him. Their eyes met in a way that had not happened before and the looks lingered.

The simple soda Brad was sipping suddenly became an elixir that intoxicated him to the point where all control was lost. Rena must have felt the same inebriation because her eyes never left his as he moved toward her and took her in his arms.

Their kiss was long and passionate. Brad could feel her body heat through her clothing as though she was not wearing anything. Soon she was not. Each explored the others body with hands and lips and eyes as they lay atop their clothing spread out beneath them.

It was all Brad could do to restrain himself when they finally became one on the roof beneath the sun. Their coupling was as natural as the earth and the sky and when they were both satisfied they held one another closely until unwanted drowsiness caused them to return to the reality of the day.

Chapter Eight: CONFRONTATION

The development of the MothBot was coming along nicely. The miniaturization of all its internal parts, the biggest obstacle, had been overcome. A few steps still needed to be taken to make it fully operational, but the team was on track to accomplish the tasks. For the present, the scientists spent much of their time enjoying the fruits of their labors. They practiced flying the MothBots inside Powervert's warehouse and away from inquisitive eyes. Brad and Rena did not participate in the piloting experiences, but they enjoyed watching the four members of the scientific team, each with his own MothBot and controller, trying to outdo one another with their flying skills. And they were becoming extremely skilful. The 'boys at play', as Rena called them, were able to command the drones to take off and land, fly in formation, and even fly backwards or inverted. All the Powervert people felt justified in enjoying a little leisure time because they had been successful enough at developing the product to warrant an eleven million dollar research grant from the U.S. Department of Defense.

Flying MothBots around the warehouse was not all play all the time. The research Powervert undertook differentiated itself from the goals of other companies with the ability to power the drones from afar by beaming electricity wirelessly from the ground to the aircraft. Powervert had long since developed a way to beam power to and from fixed objects. Now the challenge was to do the same when one or more of the objects was in motion.

Clearly, some sort of tracking device needed to be utilized to accomplish this goal. Their 'Lazigun' that sent the electrical energy, and the 'rectenna' that received it, had to be configured to both send and receive GPS location information so each element would know the exact position of the other. Each MothBot was fitted with miniaturized version of the Lazigun in the nose and a rectenna in a small spherical housing below. This way, the aircraft could be lined up to form a 'power chain' from a ground-based Lazigun to the most forward MothBot.

The tests of the system worked well when the drones were lined up in stationary positions on the warehouse floor, but when flying, the electrical signal would sometimes fail to make a connection. After an extensive amount of brainstorming, the scientific team concluded that a continuous beam of electrical energy could not be maintained because the drones moved about too much. So the electricity, instead of being sent directly to each

MothBot's motor, was sent instead to its batteries, and the Laziguns making a connection only intermittently with the rectennas, were now able to provide each MothBot with a continuous supply of power. This application, known as the 'burst theory' would take place each time the GPS coordinates of a power sending Lazigun and a power receiving rectenna locked together.

The real test of the practicality of the system came when the group moved their experiments outside. Now the MothBots would be subject to the vagaries of the wind and other natural phenomena like temperature and humidity.

The tests were accomplished by firing electricity from a Lazigun on Powervert's roof to a succession of MothBots flying in a line between the roof and David's Place 4.3 kilometers away. The farthest MothBot would be commanded to use its camera and image sending capabilities to send real time video back to the monitors on the roof to verify the test.

One scientist, usually Bratley, would be stationed at the barn to observe the last MothBot in line. Observation of the other in-line MothBots was unnecessary because each of them could use their own onboard camera to transmit videos of the next drone in front, to which they were transmitting electrical impulses.

The scientific team was surprised and pleased to discover that it took only one MothBot to send electricity to the one stationed near the barn. This meant that if five other MothBots were similarly spaced to receive and re-send the power, the total distance could be a minimum of 25.8 miles.

They set out to prove this by flying the series of MothBots in a U-shaped pattern with the observing one remaining stationary at David's Place. Then the U-shaped pattern was expanded until they had achieved an overall transmission distance of 73 miles.

When the verifying video was displayed on Powervert's rooftop monitor, the crew was ecstatic. This meant the MothBots could now be configured to government specifications and sold to the Department of Defense and the Department of Homeland Security for utilization in the battle to stop the threat of terrorism.

The team organized by the Listener's Contact was also making progress. It seemed they had already been working on the development of a similar technology, so when the MothBot was presented to them to be duplicated, much of their previous work could be utilized.

Soon the drone they now called a 'Heliorobo' was being mass-produced and distributed. It differed somewhat from the original MothBot in a number of functions. When a MothBot was deployed and receiving electrical energy to keep it operating, a Heliorobo could position itself between the MothBot and its power source, receive and use the energy being transmitted, and remain undetected. When in this position, the Heliorobo could receive the video signals being sent from the MothBot and the Heliorobo operator would know what the MothBot's operator was observing. In addition, the 'piezo poison pill' feature of the MothBot had been compromised and the Heliorobo operator could activate it at will. In other words, any MothBot within range could be made to destroy itself whenever a Heliorobo operator thought it was 'seeing' too much of their operations.

Of course all the Heliorobo's features would be wasted if the Contact's associates did not know the whereabouts of the MothBot or chain of MothBots. The solution to this problem was found when it was discovered that although the MothBot itself was virtually invisible, the device sending control signals emitted electronic pulses that were easily picked up with a simple scanner. With the scanner locked on the controller's signal, the exact location of the MothBot became known as soon as its GPS coordinates were broadcast.

Even though the replication of the MothBot came rather quickly and easily, it was accomplished only with the continued input from the Accomplice. He was distressed each time the Listener contacted him with a technical question about this or that. His desire to divorce himself from the Listener and his associates was not going well, and he was beginning to suspect it never would.

The Accomplice decided to confront the Listener hoping that the revelation of the things he knew about him would cause the Listener to fear exposure and leave him alone.

Even though the Accomplice was fearful that a confrontation might aggravate the Listener into doing something drastic, he believed the more he let the Listener know that he knew, the better his chances of intimidating him would be. He planned his confrontation to happen at the train station when the Listener was meeting with his associate.

Since the Listener and his associate met while pretending not to know one another, the Accomplice would wait until they had spoken to one another for some time before approaching. Then, hopefully, the other person would simply leave, and the Listener would recognize that the Accomplice was aware of these train station meetings.

The Accomplice arrived at the train station prior to the Listener's regular Thursday-at-four meeting. He took a seat near other travelers knowing that when the Listener came he would seek some isolated place. He had guessed correctly.

The Listener arrived at the regular time and seated himself at a distance from others. He failed to notice the Accomplice who had covered his face with the newspaper he was pretending to read.

When the Accomplice lowered the paper to see what was going on, he noticed the other person had joined the Listener. They were engaged in an animated conversation without the usual pretence of being strangers. It seemed clear to the Accomplice that the two were having some sort of confrontation and he began to wonder if he had picked an inopportune time for a confrontation of his own.

Nevertheless, after enough time had passed that the Accomplice thought to make his move, he set aside his newspaper and approached the two men. His action caused the Listener to stop speaking in mid-sentence and stare at him. His associate, noticing the distraction, stared also. The looks from these two men were so sinister; chills ran down the Accomplice's spine, but he continued his approach. He tried to force a smile but knew he was not successful as he greeted the Listener.

"Hey there, how are you?"

"I am well; and you?" the Listener answered.

"Planning a trip?" the Accomplice challenged.

"No, my friend and I are only here to talk in a quiet place. And you, where are you going?"

Things were not working out as the Accomplice had planned. The Listener's friend remained seated where he was, glaring at the Accomplice. Neither he nor the Listener made an attempt at introductions.

Now the Accomplice was searching for an answer to the Listener's question, and he knew he was taking too long to think of a response.

He finally said, "I'm not going anywhere either, I just stopped in here to get a magazine." Then he quickly added, "I must be going now. Good to see you."

The Listener only nodded as the Accomplice took his leave wondering how he could have been so stupid as to think things would automatically go the way he wanted. The intimidation he had meant to bring to bear on the Listener had backfired and he was the one who felt disadvantaged.

Rena called Brad into her office. He dreaded the thought that their intimate rooftop encounter would result in giving her a feeling of superiority over him, and here she was— demanding an audience with him.

The things the summons seemed to be on the surface, failed to materialize. When Brad entered her office she was perky and pleasant as usual. She wanted to discuss business in general and the business about Jason in particular. When she and Brad had gone over the general business issues of the day, she suggested they should do something about the suspicions surrounding Jason other than just sitting back and waiting for something to happen.

They agreed the best thing to do would be to confront him with the situation as though he was not under suspicion and observe his reactions.

After all, Jason owned as much of Powervert as each of them did, and he deserved a chance to have the suspicions about him removed, or to suffer the consequences of his disloyalty.

Rena contacted Jason over the intercom and asked him to join she and Brad in her office right away.

Jason was not doing anything that could not be set-aside for a while, and he soon popped into Rena office bursting with enthusiasm about the enviable position Powervert was in and its prospects for a bright future.

Rena hated to burst his bubble with the seriousness of the situation, but she did not hesitate to change the subject to the compromised position Powervert was in. Both she and Brad watched Jason's expressions intensely when the word 'compromised' was used. No reaction other than surprise was observed.

"Are you telling me that someone has stolen the secrets about the MothBot?" Jason demanded to know as his cheerfulness evaporated. "How is it both of you know about it and I don't?"

"It developed slowly," Rena responded, "and we weren't sure until a short time ago." Then she continued, "Have you noticed anything suspicious or anyone acting suspiciously?"

"No," Jason answered, "but before we go any farther, I need to hear how this all came about."

"Fair enough," Brad interjected. Then he began explaining the microphone and recorder at Grubby's along with the other details of the events excluding, as he did, the wiretapping of Jason's telephone and the other incidents leading to his suspicions about Jason. All the while, Brad watched Rena watching Jason for signs of guilt. She apparently did not see any, nor did Brad.

Brad decided that telling Jason the truth about his spying would only cause a rift between them, so he fabricated a story about the cleaning lady telling him that she noticed a recorder in Jason's credenza. She had said it was only because she wanted to buy one for her son that she noticed it at all.

Brad recognized the weakness of his lie when Jason wondered aloud why the cleaning lady had not asked him about it. Brad said he could not explain why she did that, but the point was that the recorder was there, and it caused him to wonder about his conversations being recorded at Grubby's.

"So, you think I'm the traitor," Jason was indignant.

"No one is accusing you of anything," Rena said. "It is just that we were wondering why you have a recorder in your office."

"This is beginning to get ridiculous," Jason was beginning to get angry. "In the first place, that box in the credenza is not a recorder."

Then his disposition began to calm down as he explained, "Well, a person can record with it I suppose, but it is actually a converter that belongs to the cable company. It's only where it is because I forgot to return it when I switched to satellite TV reception."

The others in the room must have noticed Brad's look of relief when he heard Jason's explanation. The look quickly changed to one of concern as Brad realized he had been wrong all along. But if Jason was not guilty of selling out Powervert, who was?

"Tell me, Jason, do you have any ideas about who might be stealing company secrets? It must be someone inside the company."

"I have no idea," Jason responded, "but it doesn't make sense for someone inside the company to record your conversations because each of us already knows the same things you do about the technology. One of us could simply copy the files and give or sell them to whomever he wished. I think it must be someone outside; say someone from Dronomics maybe."

"You think our business partners would do something like that?" Rena asked.

"Do you think one of us would?" Jason countered.

Brad interjected his thoughts into the discussion by saying, "It makes sense that someone outside the company, but close to it, would be in a position to gather all the technical information needed, so the only thing they wouldn't have would be the development plans and they hoped to gather these by recording my conversations."

"I agree, that sounds more reasonable than anything we've discussed before," Rena said.

"Is that because I was the one thought guilty before?" Jason speculated.

It's time to lighten this thing up, Rena thought before she said, "You, sir, are clearly guilty by association with the likes of Mr. Ganderson and myself."

Relaxed smiles and glances all around told her she had been successful before she added, "We're not finished here, but I suggest we take a 'Norman Special' break before we go on." The Norman Special was a cup of hot Arabica coffee laced with a half-ounce of Frangelico liqueur.

After Rena had prepared and served the special coffee, the trio began again in earnest to find a way to expose the industrial spy—whoever it was.

Brad explained that the chances of success were slim because the thief already had all the information needed to compromise Powervert's plans, and that the trails he followed had gone cold.

Privately, he thought about his suspicions concerning Jason. The recorder had been explained away, but what about Jason's disappearance the night the things were delivered to whoever was buying them? The other scientists had stayed at home that night—or had they? Just because their cars remained in their parking places was hardly conclusive evidence that they were at home. One of them could have used another means of transportation to deliver the goods, or even walked for that matter. And there was one other thing. Brad had not seen Gordon's car because it was inside a garage. It could have been used for the delivery and returned to the inside of the garage without Brad's knowledge. He decided to come clean with Jason about his spying.

"Jason, I have already told Rena about this, but I tapped all the office phones hoping to find our spy. The effort was rewarded with a call from someone with his voice disguised that a delivery would be made that night.

The contents to be delivered were not disclosed, but I assumed it to be something concerning the MothBot and its controller.

"At that time, I suspected anyone in the company, including you and Rena, could be the guilty party, so I monitored the nighttime activity of each member of our research team. Slade's car and Bratley's car did not move from their parking places all night. Gordon's car could not be seen inside his garage and it could have been used to make the delivery without my knowledge. Do you think there is any possibility that Gordon is the one we're looking for?"

"I don't think it's any of us," Jason did not hesitate to say. "You never mentioned my car. Did you skip me for some reason?"

Brad felt pressured to continue with his disclosure. "I saw your car was missing around 9:30 and it didn't come back 'till around midnight."

"So that led you to think of me as the bad guy."

"Yes."

"Well, I'm not the one. I don't know what night that was, but I sometimes go out to a late movie."

"After talking to you and listening to what you have to say, I'm now convinced that you had nothing to do with the things that are going on. And your suggestion about someone at Dronomics possibly being the one makes a lot of sense. But you should be able to understand how the things I observed led me to conclude the things I did."

"Don't worry, Brad, you're off the hook. But while you were talking, I was thinking. If someone at Dronomics is guilty, how could they get the information about the miniaturized insides of the MothBot?"

"They could steal a completed model," Rena offered.

"OK," Brad joined in, "but how could they do that? We're not missing any completed MothBots, are we?"

"No," Jason said, "we still have all the models we ever had … except the two that burned up."

"Then none have been stolen," Brad concluded.

Rena was the first one who guessed, "None have been stolen unless the two that burned up didn't burn up at all!"

"Ohmygod!" Brad and Jason said in unison before Brad went on, "Did you verify the destruction of those two, Jason?"

"Yes, I did, although there wasn't much left to see after the fire. Everything about the MothBot is flammable except some of the metal parts and they were melted together and distorted from the intense heat."

Brad wondered how the MothBots could be made to self-destruct. Jason reminded him about the drones being made this way on purpose, but the fact that the destruction had happened accidentally was inexplicable.

Rena continued with her speculation. "And if it wasn't an accident at all, isn't it possible one of the MothBots was burned up and made to look like they both were?"

"It sounds like we're on solid ground here," Brad said. "If one of our scientists is guilty, which one could it be?"

"You may have answered your own question," Jason observed. "Didn't you say that Gordon's car could have been used to make the delivery?"

"It's still speculation, Jason," Rena contributed. "Is there any other evidence that might point to Gordon? You should be the one to answer the question because you're around him much more than either of us."

"Nothing comes to mind. I'll try to keep tabs on him. Maybe he'll do something to trip himself up."

"I don't think waiting around for some other bad things to happen to us is such a good idea," Brad suggested. "Why don't we call Gordon and the others in here one a time and question them about things?"

Rena liked the idea of bringing the situation to a conclusion as soon as possible, but she thought they would be more successful if they had some sort of plan.

"After all, if our questions don't result in finding the guilty party and one of these guys is guilty, the bad one could simply use the information he got from us to further distance himself."

"OK," Brad summarized, "we can all watch our team of researchers for suspicious signs until we can come up with a plan for conclusive questioning. In the meantime, I think Gordon's house should be checked out to see if there is any suspicious evidence in his garage. You're the right person for that, Jason, because you can visit his place pretending to want something pertaining to the research."

"I'll do it," Jason agreed. "As a matter of fact, I'll stop by some time tonight. Who knows what I might find?"

It was shortly before 11:00 o'clock when Brad's cell phone began ringing impatiently. At first he could not remember where he left it and by the time he remembered and picked it up, it had switched to voicemail.

Brad thought about setting the cell phone back down, but decided to listen to his voicemail in case it was something important from Lissy or another of his personal VIPs. It was even more important than that.

It was a message from Jason telling Brad that he was right. Gordon confessed everything. Could Brad come to Gordon's house right away? Brad pressed the 'RESPOND' button on his cell phone.

"Jason, it's Brad. I just got your message. Is what you're saying true?"

"Yes, Brad, I'm here with Gordon now. He says it all started innocently, but now he is angry with himself because he found out the person he is giving the information to is associated with terrorists. He said he would tell us everything. You need to come over h ...."

The call dropped.

Brad had heard enough. He would go to Gordon's and listen to his confession. First, however, he searched around his apartment until he found his hand-held recorder. He did not want to miss any part of the things he was about to hear.

A feeling of foreboding swept across Brad's subconscious as he drove toward Gordon's house. He could not seem to shake the feeling that was soon replaced by one of shock and fear when he saw an orange glow in the sky directly in his path.

The worst thing he could have anticipated was happening right before his eyes. As he rounded a corner he saw Gordon's house completely ablaze. Fire Department trucks were just arriving at the scene, and his way to the house was cut off by Police cars restricting access with their flashing red and blue lights.

He got close enough, however, to see Jason's car parked in front of the house that was blazing so ferociously, no one could get near it.

A policeman who came to his car and asked for his driver's license surprised Brad.

"Is this your current address?"

"Yes," Brad replied.

"This address is on the other side of town. What are you doing in this neighborhood?"

Brad was not sure how to answer. He hesitated some time before saying that one of his employees lived in the burning house and that he was only on his way for a visit.

The officer left Brad alone after he explained his association with Gordon and had identified the company they both worked for, but the slow response on Brad's part made the officer suspect that he could be hiding something by making his story up as he went along.

Chapter Nine: VENGEANCE

Beginning the moment the Listener observed Gordon spying on him, he prepared a defensive shield by moving into a new residence and arranging to have listening devices placed in Gordon's home. He was not sure what the consequences would be if he were found out, but he knew that his pipeline to the wealth he was to receive from the Contact's people would be broken. This would be unacceptable. He would have to devise a way to rid himself of Gordon if his tongue became too loose.

When the Listener was given a dozen Heliorobos, it was a sign that the Contact had been true to his word and the promised money would be forthcoming.

The Listener wasted no time familiarizing himself with the new drone. His practice with the MothBot had been extensive enough that soon he was able to master the operations of this newest incarnation. But the Listener's ability to command the Heliorobo to fly around as he wished was only part of the various features of the drone to be mastered.

Knowing which routes could be safely flown without crashing into an unseen wire or other obstruction was vital to the success of any flying mission. Sending it to a specific location could be accomplished with the onboard GPS only if the flight path was obstacle-free.

The Heliorobo's digital video equipment, set for normal daylight or nighttime infrared, would show the flight controller the objects the camera saw in real-time. This enabled the operator to pilot one drone as though seated in the cockpit. When more than one was aloft, the operator was faced with more complex operational challenges.

Energizing a drone's batteries was done automatically by sending an electrical impulse via a Lazigun to the craft's rectenna. On the other hand, operating the Lazigun aboard each Heliorobo to fire a charge of electrical current ahead to a target was something that had to be mastered by the operator.

And the last major feature, the one the Listener liked best, was the ability to destroy the Heliorobo by commanding the piezo poison pill (PPP) it carried to ignite. The Listener thought to improve this deadly feature by including a packet of jagged pellets of magnesium inside each drone. The irregular shape and the jagged edges made the magnesium easy to ignite, and once set ablaze, the fire consuming the pellets could not be extinguished.

When the command to self-destruct was sent to a Heliorobo, the flames resulting from the burning hydrogen would destroy the drone and ignite the magnesium, which would burn slowly and extremely hot for a long period of time. It was not long before only a few people on the planet could operate these deadly devices as efficiently as the Listener.

The Listener's anticipation that Gordon might try to disclose his identity proved to be prophetic. Only a short time after having installed the listening devices in Gordon's home, the Listener overheard a conversation that could have led to his exposure.

Gordon was telling someone else about his betrayal of his company between sniffles and sobs. He had clearly cracked. The Listener's name was not mentioned because Gordon did not know it, but the confession would surely lead to an investigation. Something had to be done quickly.

The semi-enclosed patio behind the Listener's apartment provided an ideal place to conduct the operation he planned. He set one of the Heliorobos on a coffee table, pointed it in the direction of Gordon's house, and attached wires to it that were plugged into an exterior convenience outlet. This allowed the Lazigun mounted in the drone's nose to beam forward an electrical current.

Next, he selected Heliorobo #2 and commanded it to fly to the GPS coordinates of Gordon's house and remain stationary at an altitude of 100 feet.

Heliorobo #3 was then selected on the controller and commanded to fly to a destination half way between the two residences and maintain a fixed position and altitude of 300 feet with its tracking system locked on H2.

The Listener was pleased about being able to select one or more of his drones with the same controller. He switched to H4, commanded it to lock onto H3, and maintain a distance of 500 feet. H5 was locked likewise onto H4 and H6 to H5.

This way, all the drones were linked together in a chain that went fromH1 on the Listener's patio to H6 then H5, H4, and H3 before reaching H2 at Gordon's house.

With the drones planned for this operation in place, the Listener locked H1 onto H6 and activated the camera in H2.

The high-definition monitor the Listener watched displayed Gordon's home clearly. Even though it was after dark, the infrared camera was not necessary because the lights from the house provided all the detail the camera needed.

The Listener maneuvered H2 around the house until he saw two men talking in the family room. It was better than he had hoped. The sliding glass door was wide open. Not that it mattered to the Listener; he would have crashed the drone through the glass without fear of damage due to the strength of the craft's construction if it had been necessary.

The Listener could see fear in the eyes of both men when they realized the Heliorobo was hovering near them. He activated H2's Lazigun and fired it into Gordon's chest. Quickly turning, H2 fired again into the chest of Jason before he could react. Both men crumpled to the floor.

The noise of their falling must have alerted Gordon's wife because she came rushing into the room. The Listener could see the horror on her face as he carefully aimed H2 and fired its Lazigun.

The urgency to obliterate all traces leading to him was now magnified as the Listener followed through with the completion of his plan. He commanded H3 to trace the GPS path created by H2. Since the other drones were set to maintain a certain distance from each other, they all followed H3 as it headed directly for the Accomplice's house.

The Listener activated the onboard camera of H3 and commanded it to stop when it had reached Gordon's house. With five Heliorobos hovering in sequence, the Listener began the final phase of his deadly attack.

The Heliorobos were unlocked from one another, and one by one, each onboard camera was used to send each drone to its final destination.

Heliorobo # 2 was sent into the kitchen where it was made to crash into the wall behind the stove where its self-destruct feature was activated. Number 3 was flown into the living room and self-destructed near an electrical outlet. All the other drones, except H6, were made to do likewise in other parts of the house—all near electrical appliances or outlets.

The Listener used the camera aboard H6 to monitor the effect of his work. It was more spectacular than he had anticipated. The flames from the hydrogen followed by the intense heat produced by the burning magnesium pellets created a fire that engulfed the house and garage quickly and completely.

The drone was repositioned a distance from the blaze. It could easily be destroyed if it got too close to the heat rising from the fire or if the downdraft from a police helicopter sent it crashing to the ground.

The Listener watched the scene develop as the police began to cordon off the area and the fire trucks arrived to spray water on the neighbor's houses. Then he saw something else.

Brad's car came up the street only to be stopped by a policeman. To verify it was indeed Brad in the car, the Listener maneuvered the Heliorobo to within a few feet of the car's windshield. Yes, it was Brad and he was giving the officer his driver's license.

The Listener wanted desperately to send a bolt of electricity into Brad, but it was not possible because the chain between Heliorobo #1 and H6 had been broken. He could not resist, however, sending the drone close enough for both Brad and the policeman to see.

The officer saw Brad's eyes shift, and then stare at something else. As he turned to see what Brad was looking at, H6 was commanded to zip away at top speed. The policeman was not sure if he had seen something or if he was hallucinating.

The Listener had achieved his goal. He associated Brad with the fire and the Heliorobo through an eyewitness. The eyewitness was also a cop. The stage was now set. All the Listener had to do was to make sure Brad was blamed for the fire and the deaths.

Chapter Ten: THE VOICE

Powervert's office the next morning was a scene of distress. The company had lost one of its owners and one of the scientists. The morning paper only said "The three victims claimed by the tragic fire were burned beyond recognition, but believed to be the owner of the house and his wife along with another adult male." Brad, however, knew them to be Gordon and his wife along with Jason Grilling.

Rena was not dealing with the situation lightly. She and Brad and Jason had come to be the owners of Powervert when Norman, the previous owner suddenly died. Then David, their chief scientist, was tragically killed in a fire in the company barn. Now two other Powervert employees had their lives taken from them in yet another fire. Rena was beginning to wonder if there could be some sort of curse that had been cast on her and those around her.

"Listen, Brad, I'm not sure I'm going to be able to handle all this," Rena said as tears began welling in her eyes.

Brad took her in his arms and held her closely without speaking. The tragedy that had befallen them was affecting him as well. They stood there for several moments holding one another before Brad realized he was feeling sorry for himself and should take charge of the situation instead by doing something about it.

He gently eased Rena away with his hands to her shoulders and looked directly into her eyes. The thought of her beauty and desirability flashed across his mind before he forced himself to be pragmatic.

"We can regret the bad things that have happened to us, but we mustn't use this as an excuse to stop fighting back. These things are not happening in a vacuum. Someone is orchestrating all the things that are occurring for whatever reason."

"Why do you say that, Brad, do you think someone set the fire on purpose?"

"Think about it, Rena. Jason went to Gordon's house because Gordon was ready to spill his guts and he told Jason everything. Jason called me over to witness it and when I got there the fire was well underway. It could have been a coincidence that whoever started the fire did it just then, but it seems more likely that someone knew Gordon was confessing everything and decided to silence him forever."

"So you think the person who burned down the house was the one Gordon was selling the technology to."

"I'm convinced of it. And there's something else. I saw a MothBot hovering near the fire."

"Oh, my God!" Rena cried. "One of our drones near the fire could implicate us."

"I'm not sure it was one of ours. It looked a little odd, but it was definitely a MothBot type. Why don't you make a couple of our Frangelico coffees, and we can sit down and try to reason this thing out."

They sat in Powervert's reception area; Rena behind the desk and Brad in a guest's chair. For a while they only sat and sipped their liqueur-laced coffee. The calming effect on Rena was noticeable and Brad was thankful for it. Cool clear heads were needed if any mysteries were going to be solved.

Rena was the first to offer a synopsis of the present situation.

"We have a plan to build a spy drone. Someone finds out about it and works out a deal with Gordon to get information. Gordon sells the tech plans to this 'someone' and then regrets it. He feels guilty and confesses to Jason. The 'someone' finds out about the confession and burns down Gordon's house killing him and the others. Telling someone about someone else's plans, or even selling a working model to them hardly provides a reason to murder anyone … let alone to kill three people."

"I agree," Brad answered, "so let's take your points one at a time and search for any justification for murder.

"One person wanted to tap into our tech advances and that person was not Gordon. Gordon already had all the information any of us had. That person … let's call him … or her … the 'Culprit' … must have been spying on me before Gordon got involved. Gordon would not have had to put a microphone under the table at Grubby's because he had direct access to all our plans. The Culprit then solicited Gordon for additional information about our plans to build a spy drone."

"No reason yet to kill someone," Rena observed.

"Then the Culprit discovered the deadly aspects of the drone that we developed only for the purpose of gathering information. By 'deadly aspects' I mean the MothBot's ability to fire electrical impulses for one, and destroy itself whenever it was needed for another.

"No sinister reasons yet, but the next point you brought up might provide a clue. I'm talking about Gordon's regrets about selling the technology. Gordon could have had a change of heart about selling out his company, but more likely, he regretted giving the information to someone who was going to use it for evil purposes."

"I still don't see how murder could be warranted," Rena said, "unless the Culprit only did it to keep Gordon from identifying him."

"OK," Brad said, "let's run with that. If the Culprit wanted to remain unidentified … and was willing to kill people to make it happen … his identity must be hiding his own personal guilt of some crime, or hiding the path to others who are criminals. Otherwise, it wouldn't make sense to remain unknown. Industrial espionage is against the law, but it hardly provides a basis for murder. So let's say the Culprit is someone wanted by the authorities. When he realized his identity was about to be disclosed to Jason, the Culprit set Gordon's house afire; killing the three people. Now we have all the more reason to find out who this person is. Not only did the Culprit do this horrible thing, whoever it is has stolen our proprietary property and is using it against us."

"What do you mean by that?" Rena asked.

"I already told you that I saw what looked like a MothBot near the fire. And you already know about its self-destructing capabilities. I believe the Culprit used one or more MothBots to start the Gordon fire, but this conclusion begs the question: Why didn't Jason and the Gordons run out of the burning house?"

Rena was quick to respond. "They couldn't run out if the Culprit killed them first, then set the fire to cover up the crime."

"Bingo!" Brad agreed. "An autopsy should reveal the cause of death and that could verify our assumptions."

"I think it's time to go to the police and tell them what we know," Rena suggested.

"I'm not sure that's such a good idea," Brad countered. "The police probably already know as much, or more than we do ... and there's another thing. A policeman at the house saw me, and I work with the people that were killed in a fire that was started by a drone that our company developed. Surely, they would suspect me of being at least partly responsible with nothing more than circumstantial evidence."

"So you think we should try to find the Culprit ourselves?"

"Yes."

It was quite by accident that Brad noticed the small article on the fourth page of the local newspaper.

> MEDLINTON – An autopsy has been completed on the victims of the tragic fire.
>
> Even though each person was burned beyond recognition, they were identified as Mr. and Mrs. Gordon, the residents of the home, and one Jason Grilling, also a resident of Medlinton.
>
> The cause of the deaths remains undetermined because the victims were found to have no toxic substances in their bodies.
>
> Preliminary evidence indicates that each died from natural causes. Since it is improbable that three people would die naturally at the same time, the authorities suspect foul play.
>
> The case is under investigation.

The words hit Brad like a sledgehammer. He remembered the demonstration that Derek gave that afternoon so long ago in the Powervert warehouse. The MothBot had hovered close to each of the observers, pointing its nose directly at one individual at a time.

If a Lazigun had been installed in a drone, and had been charged with electricity, it could have easily sent a deadly bolt into each person selected.

Now no doubt lingered in Brad's mind that the victims had been murdered by electrocution before being partially cremated. This Culprit, whoever he was, had to be found and stopped before this murdering madness could continue.

Any trail leading to the Culprit had been obliterated with the death of Gordon. The only thing Brad could think to do was to go through Gordon's files with a fine-toothed comb—hoping to find something incriminating.

Brad entered the laboratory when the others had left Powervert for the evening. He remembered which of the three desks the scientists used was Gordon's and he began searching the drawers. There were lots of papers to be sure, but they all contained information about the MothBot project and little else. The telephone was another matter.

Brad knew that each telephone in Powervert's system had voicemail capabilities. He also knew that in order to access one's voicemail, a password had to be punched in on the keypad. He would have no way of knowing Gordon's password, but he had an idea that might work. He picked up the handset and selected #5 to access the messages. When the recorded voice asked him to punch in the password, he selected #0. The recorded voice returned immediately and prompted him to listen carefully to the three options because the menu had recently changed. The third option was the one he had hoped for. The voice said, "Press or say 'three' if you wish to edit your voicemail." Brad pressed #3 and found the next series of options to contain, "Select or say 'one' to change your password."

Brad could only hope that changing the password would not delete any messages left in the system. It did not. When he began the routine anew, selected #5, and entered his new password, the voice said there were two un-played messages and began relaying them.

Both turned out to be calls that would only have been important to Gordon if he had been around to hear them. Brad was disappointed that his final opportunity to find the Culprit was coming to an end.

Just as he was about to hang up the handset, the voice droned out some new options. He could replay the messages, delete the messages, or play messages that had been previously deleted. When he opted to replay the deleted messages, several were heard before one began that got all his attention.

Your request for disassociation is understood but cannot be granted. The journey taken so far has brought to light all the peripherals, but not the product itself. Please see to it that a full sized working model is delivered to me before the next ten days have passed. Please do not burden me with any difficulties you may encounter fulfilling this stipulation.

Brad quickly associated this message with the one he had intercepted through his electronic eavesdropping. In this one, the Culprit was telling Gordon to deliver a working model of the MothBot—or else. And the message his snooping had produced was Gordon verifying the delivery although his voice had been disguised in a way that made it unrecognizable. That was not the case with this message, but there was something else. Brad recognized the voice!

The problem was that Brad knew the voice but he could not associate it with a face. It was the voice of a man from his past that rang in his ears like some grotesque form of torture. And it was not only the words. It was the menacing implications it conveyed that disassociated it from the other voicemail messages. In addition, the voice had demanded compliance with its wishes in a way only polite enough to substantiate its authority. The man behind the voice was clearly someone who was accustomed to being in a position of dominance.

Brad needed someone to talk to. He wanted to talk the situation over with Lissy. Rena might have been a more sensible choice, but Brad wanted to confide in someone whose perspective was not tainted by being too close to the situation. Besides, he and Lissy had been through other problem solving situations before and the results had been twofold by leading them to the correct answers and bringing them closer together at the same time.

Lissy was doing her regular job behind the bar at Grubby's Tavern when Brad took his favorite seat at his regular table. Before Lissy could greet him with a cold Coors and a frosty mug, he searched under the table with his hand until he located the microphone that had been planted there. With a firm grip and a twisting tug, the microphone was wrenched from its hiding place. Brad took a quick look at it before relocating it to the inside of one of his pockets.

"Hi, Stranger," Lissy smiled at him.

"Hi yourself, Good Lookin'," Brad grinned back. "I understand this is the place where 'Lissy the Spy' works. Do you know her?"

"Actually she's my twin sister and I'm only filling in while she's on her break. She's due back soon."

"Well, when she comes back tell her that the 'sleuth from her youth' is here with a mystery that needs solving."

"The sleuth from her youth?" Lissy could not help chuckling.

"Yes," Brad answered pan-faced. "We, along with Sherlock and Mrs. Marple, have solved many mysteries together."

"I understand the 'sleuth' part, but what about the 'youth' part?" Lissy wondered.

"It's only a private matter between she and I that she is but a youth and I am by far her senior, but we manage to make a great team with my worldly knowledge and her youthful exuberance."

The next time Lissy came to Brad's table, she came as herself expressing how refreshed she was after her break.

"I understand you have a mystery to solve," she remarked.

"Actually, it's several mysteries rolled into one and it's rather complex. Can we discuss it over a late supper tonight after you get off work?"

"OK," Lissy agreed. "Is there anything I should be thinking about in the meantime?"

"Sure," Brad offered. "Think about the serious crimes a man might have committed to justify murdering people in order to hide his identity."

Japanese lanterns around the pond made the shimmering water sparkle with their golden reflections. The scene inside the restaurant was no less romantic with the overhead lights dimmed to accent the candles glowing on each table. The couple at a secluded table near the window enhanced the scene as they sat facing each other with their wine glasses raised in a toast. The intimacy that the pair seemed to be enjoying would have to wait until later when they were alone. The romantic atmosphere surrounding them was deceptive when taken in context with the conversation the two were having.

"The point is," Brad was saying, "someone … some man … is probably guilty of murdering three people and the only reason I can think of for such an action is that the person feared being identified. Gordon knew him and told Jason about him and they were both about to tell me when they were killed."

"How do you know it was a man?" Lissy asked.

"I heard his voice on a telephone message he left for Gordon. And, as a matter of fact, I recognized it, but I can't seem to be able to place it."

"OK," Lissy said, "you asked me before why this man would want to kill people and maybe to hide his identity is one reason, but I thought of some others. Sometimes people kill people because they are insane."

"This guy is not insane," Brad countered. "At least these murders were not a random act of craziness. This is all part of a larger scheme involving stealing the secrets to our spy drone."

"So, he isn't nuts," Lissy added. "Maybe his reasons were to keep anyone from learning why he stole the spy drone plans. I mean if he took the plans so he could sell the information to someone else, say a terrorist, wouldn't that make him want to stay unknown?"

"That could be it," Brad agreed. "If he was hiding a larger scheme like supporting a terrorist organization, his association with them could lead to serious criminal charges like espionage or even sedition. And now things are beginning to add up. This guy I've already named the 'Culprit' has done a lot more than steal plans. He has developed them to the point where the drones can be used to shoot enough electricity to electrocute people and maybe start fires when they are commanded to self-destruct. And that leads to something else.

"The fire at the Gordon house was started in several places at the same time. The police think an electrical surge caused the fire because it originated near electrical outlets in different parts of the house. Using a number of drones and igniting them near electrical outlets could just as easily start fires in different parts of the house. And that means that the Culprit had access to several drones and their controllers at the same time. One person cannot build several drones in the amount of time the Culprit has had the plans. It had to be a highly sophisticated organization with state-of-the-art equipment. If we are guessing right, a terrorist organization supported by a renegade government could be behind this whole thing."

"All right then," Lissy said. "Let's just assume these guesses are the correct ones and the motive for murder committed by this person you call the Culprit is established. We are not going to be able to get anywhere unless we can identify the Culprit and have him lead us to the other people involved. In my job at Grubby's I've met practically everyone in town and I'm pretty good with telephone voices. Is there any way I can listen to this person's voice?"

"I hadn't thought about it, but there could be a way. Let's give it a try right now." Brad took the iPhone out of his pocket and punched the number for Powervert. When the rotary answering system responded to his call, he selected Gordon's extension. Gordon's recorded voice declared that he was not presently available and suggested the caller should leave a message after the tone. When the tone sounded, Brad punched #9 and the system began to play back previously recorded messages. When the message from the Culprit began, Brad handed the iPhone to Lissy.

Intensity sharpened Lissy's facial features as she listened to the voice on the phone. Even this look of concentration could not diminish her loveliness Brad was thinking while he waited for her reaction. He knew the message was over when Lissy's eyes assumed a blank stare.

He recognized something had stirred her, but he did not want to interrupt her thoughts, so he waited patiently until her eyes began to re-focus on her surroundings.

As though she was emerging from a trance, Lissy's eyes met Brad's as she whispered, "Yes, I know that voice. It's Sam Murphy!"

Brad could not believe it. Yes, the voice did sound like Sam Murphy's, but Sam had been killed in that horrible accident when his Carver crashed into the side of the old tanker *Lucifer*. Brad had searched the area carefully after the blast that sent the *Lucifer* to the bottom along with Sam and his boat. Only some floating debris was left from the explosion that in all likelihood had blown Sam Murphy to pieces. Surely, Lissy was mistaken with her identification—and yet.

"I can't believe it," Brad muttered aloud. "But if Murphy managed to survive that wreck somehow, the things that have been happening are beginning to make sense."

"He always hated you," Lissy remarked. "He even tried to kill you just before his boat ran into the side of the *Lucifer* and we thought he had killed himself instead."

"That's true," Brad agreed. "It's also true that he had inside knowledge about Powervert that he could have used to do the things he did including stealing the MothBot. And his association with Middle Easterners could provide the way he was able to duplicate the drone so quickly. This could mean that the MothBot is already in the hands of terrorists. If all this is true and Murphy is really the one behind all this, it's hard to understand why he wouldn't have simply sent one of the MothBots on a mission to kill me."

"Well, so far, he seems to be getting away with murder," Lissy offered. "The police don't seem to have an idea how the people in the Gordon fire died, so they have no reason to charge anyone. Besides, Murphy was assumed dead in the *Lucifer* explosion and that means he is able to roam around free and do whatever he wants. If he decided to murder you, it might lead to his exposure and that should be reason enough for him to find another way to rid himself of you."

"And that way could be to somehow link me with terrorists through my involvement with the development of the MothBot. If he could somehow manage to establish that link, he could also link me to the fire at Gordon's place and the murders that took place there."

Brad was animated. He felt deep down inside that they were on the right path and he searched his brain to come to a decision as to what to do. Brad hated metaphors, but they continued reappearing in the shadows of his mind. *It takes a thief to catch a thief* and *Fight fire with fire*, prompted him to learn as much as he could, as soon as he could, about the operation and handling of the MothBot as a weapon.

As serious and intense as Brad's conversation with Lissy had been, he forced the thoughts from his mind as he said to Lissy's eyes, "We seem to have earned our stars as detectives, and I think we are wasting a perfectly good romantic atmosphere. Can we just talk about you and me for a while?"

Chapter Eleven: COUNTERPLOT

Brad's meeting with Slade and Bratley, Powervert's two remaining scientists, was fruitful. As concerned as they were about the loss of their team members Jason and Gordon, they agreed to help Brad as much as they could while continuing the development of the MothBot. Brad had not been entirely candid with them about his reasons for the accelerated participation for fear they might decide their jobs were too dangerous and leave their employment. Instead of telling them he wanted to counter any technological development made by Murphy, he told them that he only wanted to help fill the void created by Jason's and Gordon's absence. They welcomed his participation.

Now Brad regretted not joining in the fun of playing with the MothBot in the warehouse when the others did. Soon, however, he began to grasp some of the nuances of making the drone do as he wished. He practiced much of the time after working hours when the others had left for the evening to avoid arousing suspicions as to why he was pursuing the task so vigorously.

Slade and Bratley would not know that inside the warehouse each night he was beginning to master the operation of two, three, and four MothBots at the same time.

Brad's increasing knowledge about the MothBot and its operation could be justified by his position as an officer of Powervert learning about his product. Beyond that, he was aware that Murphy's possession of the technology—and perhaps his involvement in the murders of Jason and Gordon's family—provided the tools to strike a deadly blow to Brad at any time. He became determined to find a way to detect the presence of a MothBot, controlled by a madman, seeking to destroy him.

Since the MothBot was practically invisible to the naked eye and undetectable by radar, Brad had to find another way to discern its presence. What could it be? He decided to consult with Bratley and Slade. After all, this knowledge would be important in the general development of the MothBot system.

"That's easy," Slade said simply. "The drone both sends and receives radio signals. All a person has to do is to know the frequency being used and listen to the broadcasts. Unless the signals have been encrypted, they will tell the coordinates of the drone as well as its altitude. For that matter, with the right equipment, one can intercept the video being broadcast by the drone and be able to see what it sees."

"Are the signals we send encrypted?" Brad asked.

"No, they aren't," Bratley answered. "Although encryption is done by simply interfacing the broadcasts with the right system, we haven't felt any need for it while our MothBot is still in the development stage."

"Don't you guys think it's about time to add it?" Brad asked. "The Department of Defense will surely demand some way to keep others from detecting the MothBot's activities."

"We'll get on it right away," Slade offered.

"And set it up so the controller-operator can change the broadcast from regular to encrypted with the flick of a switch," Brad directed.

With the scientific team headed in the direction Brad had suggested, he once again thought of the situation and the possible consequences of coming head to head with Murphy. If the things he and Lissy had concluded were correct, Murphy would try to implicate Brad somehow by associating him with the MothBot, and with terrorism. The MothBot would certainly be used as the vehicle to accomplish this, and Brad was comfortable with the knowledge that scanning for radio signals and observing the geographical coordinates of its location could easily detect the presence of a drone.

Since the MothBot's controller had not yet been modified to encrypt radio signals, the presence of the duplicate stolen by Murphy should be likewise easily detectable However, whoever it was that helped Murphy reproduce the drones and controllers could have anticipated the need for encryption and incorporated the capability into their cloned package. If this were the case, Brad would not be able to tell if a drone was nearby and Murphy could do with him as he wished.

Brad made his third trip to the spy shop. Once again, he was pleased with the device the store employee told him he needed. A scanner was available that came with a miniature mast surrounded by four antennas. The scanner searched continually for a radio signal until it would lock onto one detected to be close according to its signal strength. Then the four antennas would come into play by causing a compass arrow to point to the location of the signal. The scanner suited Brad's needs perfectly because it would point to any nearby drone even if the radio signals being broadcast were encrypted. One of the three he purchased was set up in his office, one in his home, and the other in his car. And since the approach of any drone was certain to be from outdoors, Brad made it a point to carefully monitor each respective scanner before he ventured outside a building or got out of his car.

Now, Brad believed he was ready for any encounter that Murphy cared to initiate. He had acquired six MothBots under the pretence of researching their capabilities and limitations. He kept two in his office, two in his apartment, and two in his car along with the controller and video monitor for each pair.

After the six MothBots came into his possession, Brad made modifications to them about which he beamed with pride at his inventive cleverness, although it was rather basic. He had placed beside the Lazigun in the nose of each one a thin shaft of metal that protruded a short distance from the front of the drone. The function of the shaft was to cause the piezo to strike a spark when the drone hit an object and forced the shaft backward.

Brad had turned the MothBot into a weapon of destruction because it could be sent crashing into a person or anything else and the shaft would strike the piezo causing the drone to explode in a ball of fire.

Of course, the PPP feature of the MothBot had always been one of its major attributes. And now, the drone could be commanded to self-destruct with a switch on the controller, or it could be purposely crashed into something with the same result.

Brad's plan was clear. If an alien MothBot was detected and its coordinates and altitude were known, Brad would program one of his MothBots to go to the precise location of the alien one. Since two objects cannot occupy the same space at the same time, Brad's MothBot would crash into the alien one resulting in the destruction of both.

There would be no need to leave the collision to chance alone. Brad could utilize the MothBot's video to facilitate the location of the alien drone visually, and then control the MothBot's flight path to strike it.

If Murphy was going to be successful murdering Brad, he would have to think of another way.

Sam Murphy had been thinking of murdering Brad for a long time and he had thought to do it with one of his Heliorobos. However, for reasons Murphy was not even aware of, simply killing Brad was not good enough. He wanted more. He wanted to see Brad suffer as Brad had made him suffer. He wanted to see Brad as an outcast of society just as he had been made an outcast. Yes, he wanted Brad to die all right, but it had to be slow and painful and in a way that would be disgraceful.

Murphy went to a pay telephone wearing transparent plastic gloves and removed the handset from its cradle. With a piece of cloth over the mouthpiece to disguise his voice, he said to the person who answered the 9-1-1 call, "Brad Ganderson killed three people at the Gordon house because they were onto his plans to sell his destructive invention to terrorists."

"Identify yourself," the operator demanded.

Instead, Murphy continued speaking; knowing that everything said would be recorded.

"The murders were done with his invention called the MothBot." And with that, he replaced the phone on its cradle and left the booth.

Murphy was aware of the newspaper article that reported the fire and the deaths were under investigation. His message would surely cause the police to investigate. And they did so quickly, but only after they had obtained a search warrant for Brad's apartment and his place of work.

Detective Nettbron was polite and courteous as he sat in front of Brad's desk. It was difficult for him to think of the executive before him as someone who would murder people and aid terrorism, but he had his job to do and looks were sometimes deceiving. He got right to the point.

"I'm here, Mr. Ganderson, to investigate the suspicious fire at the Gordon residence and I'll be honest with you. We don't know how those people died and we don't know how the fire started. Our hope is that you might be able to enlighten us because both those men worked for your company. Do you have any knowledge of the circumstances that might provide us with some answers?"

Brad was intimidated but he tried to remain calm.

"I'm sure you know more about the incident than I do," Brad responded. "But I can tell you this; the loss of Jason and Gordon was a greater blow to our company than you can imagine. The products being developed by those scientists were, and are, the life-blood of this company not to mention losing two respected and trusted friends."

"Can you tell me the nature of the product they were developing?" Nettbron asked.

"I can only say this; they were working on a top-secret project for the Department of Defense."

"Would the name of this top-secret project be MothBot?"

Brad was shocked. *How could Nettbron know the name of Powervert's drone?* On second thought, the name had been discussed openly even though the project was supposed to be veiled in secrecy, and having this inside knowledge was probably the least a person might expect from a detective like Nettbron.

"Yes it is," Brad said simply.

"And would it be correct to assume the project is still underway?" Nettbron continued.

"It is."

Nettbron, sensing Brad's trepidation, decided to play the 'good cop, bad cop' game and since he had come to Powervert alone, the 'good cop' role seemed the most suitable. He began by trying to gain Brad's confidence.

"As I said, the reason I'm here is because of the suspicious nature of the fire at the Gordon's. We think the fire could have been set intentionally because it started in more than one place at the same time. We can't be sure about this because it's possible for fires to start this way if there is some sort of electrical surge in the house's wiring system.

"One thing we can be sure of is that three human beings lost their lives in that house. It's possible that their deaths and the fire could have been caused without outside involvement however unlikely that seems. And that's the reason I'm here. I want you to tell me what you know about the case that could lead to the answers to some of these questions."

The look on Brad's face told Nettbron that he understood the things being said, yet there was no indication of a defensive reaction.

Brad sensed that Nettbron would not give up until he had heard enough about Powervert and the MothBot to be satisfied. Besides, Nettbron did not pose any threat; he was simply trying to do his job and solve a mystery. Brad wanted to solve the mystery as well.

"I'll do what I can to help," Brad began. "I agree with what you said about suspicious circumstances and there's a reason for me to think this way. For some time, my associates and I have though that someone out there could be attempting to steal our secrets and we think that Gordon could have had something to do with it."

Brad realized he was withholding a lot of information, but he believed the less this detective knew about Powervert and its operations, the better off everyone would be. He continued with his veiled revelation.

"Jason, one of the company owners who died in the fire, went to Gordon's house that night because he wanted to find out if Gordon was involved in the industrial espionage we suspected was going on.

"Apparently, Gordon knew a lot because Jason called me to come to the house to listen to the things Gordon had to say. By the time I got there, the fire was well under way and the police and the fire department were on the scene.

"Unfortunately, whatever Gordon was going to tell me and Jason was never known, but I suspect he was working with some culprit to steal our plans and sell them to terrorists."

Nettbron could not understand the strange look that flashed across Brad's face when he said 'culprit' and he suspected Brad was not being totally candid with the things he was saying. Nettbron, however, went on with his query as though nothing unusual had been noticed.

"So you think these 'terrorists' could have managed to murder your people and burn the house to protect their identity?"

"That's precisely what I think," Brad acknowledged.

"And I suppose you have a theory about the way the crime could have been committed without leaving incriminating evidence," Nettbron challenged.

Brad was torn between using this detective to help find Sam Murphy and his instinct to keep the things he knew to himself. He was beginning to develop a level of trust with Nettbron, so his desire to bring Murphy to justice prevailed.

"It's possible that our MothBot could have done the job."

There, he had said it, and now he had no choice except to reveal the MothBot's capabilities in order to explain his suspicions. Nettbron sat and listened intently as Brad expressed his reasons for putting the puzzle together the way he had.

"Our MothBot is a drone that is being developed to spy on the activities of the enemies of our country; namely terrorists. It is very small and virtually invisible as it goes about using its sensors to gather information. It gets its power from regular electricity that is beamed to its batteries so it can operate continuously without interruption. The electrical power is sent to the drone from a station on the ground and the drone has the capability to transfer the current it receives to another drone to extend the range of the surveillance. In other words, each drone can receive and send electrical charges."

"And I gather one or two persons can make the thing work from the ground," Nettbron interjected.

"Yes, that's true," Brad confirmed. "The person at the controls can see everything the drone sees with its video equipment, and the controller operator is able to command it to do as he or she wishes."

"So you think someone, somehow, commanded one of these drones to commit the crime?"

"Let's put it this way," Brad amended. "Since the drone can both receive and send electricity, it could have received enough electrical power to send a bolt of electricity into each of the victims at the Gordon's' house. I don't know if it could shoot enough electricity to kill a person directly, but it could probably knock them out long enough for the fire to do its damage."

"OK, supposing the people were incapacitated that way, what could have been done to start the fire so efficiently?" Nettbron asked.

"I'm not sure," Brad responded. "But there's another feature of the MothBot you should be aware of."

"And that is?"

"The MothBot is filled with hydrogen and it can be made to self-destruct."

Nettbron was animated. "Do you mean 'self-destruct' as in burn up?"

"Exactly," Brad concurred.

"That's interesting, but if a drone had been used to start the fire, we would have found evidence of the wreckage," Nettbron guessed.

"Probably not," Brad corrected, "the MothBot is made from carbon fiber covered with plastic film. A fire would burn up everything except maybe some hardware that could only be seen under a microscope."

"Wow! It looks like this case needs a little more investigation," the grateful detective expressed in parting to an even more grateful Brad Ganderson.

When Detective Nettbron excused himself and left Powervert, he was anxious to return to headquarters where the chief would be interested to hear about the new developments in the case. First, however, he called the police laboratory and asked a technician to go immediately to the Gordon fire and vacuum the floor at each place where the fire had started. Everything collected in the vacuum bag was to be examined under a microscope and retained as evidence if necessary. Then he drove directly to Brad's apartment. The search warrant issued by the judge had allowed him to employ the services of the security company Brad subscribed to. One of its agents met

Nettbron there and together they entered Brad's apartment and disarmed the alarm system.

There, right on the coffee table was a strange looking thing that could be nothing other than a MothBot with its controller alongside. Nettbron examined it as closely as he could without touching it or disturbing anything around it.

Satisfied that he had what he needed, Nettbron took a number of photos of the room and its contents before he and the security agent re-activated the alarm system and left the premises.

Once Nettbron reached headquarters, he went directly to his office where he loaded the memory stick containing the camera's digital images into his computer. One by one, he examined each photo before selecting several to be printed so they could be shown to the chief. Then he visited the chief's office.

Nettbron was pleased with himself. He had gotten the information he wanted out of Brad. He had taken the pictures he needed at Brad's office and at Brad's apartment without his knowledge. And he had gotten authority from the chief to continue along the path he was taking. The crowning glory of the day, however, had come when the police lab tech called and told him that some of the evidence gathered at the Gordon fire appeared to be microscopic mechanical parts. Nettbron's stomach told him it was time for dinner. His dedication to duty told him it was time to make his next move.

He asked a uniformed police officer to ride along with him as he drove to Brad's apartment.

"This guy thinks he's pretty clever," Nettbron said to the officer as they drove. "He told me exactly how he did it, and then thought I would be stupid enough to believe his story that it was done by some mysterious terrorist."

The look on Brad's face was one of surprise and concern when Nettbron accompanied by the uniformed officer knocked on the door.

"Mr. Ganderson, you are under arrest on suspicion of murder. You have the right to remain silent. Anything you say can, and will be used against you in a court of law."

Chapter Twelve: THE INMATE

Rena thought it was out of character for Silas Berthaby, Powervert's lawyer, to be pulling this prank on her. He had always been straight-laced and down-to-business and now he was telling her a preposterous story about Brad being locked up in jail.

"He needs your help," Berthaby was saying. "It happened so suddenly last night that he didn't have time to do anything until he called me this morning."

"Are you serious?" was the only thing Rena could think to say.

"I'm afraid so," Berthaby continued. "Apparently the police think he is involved somehow in the deaths that occurred during the fire at Mr. Gordon's house. Brad wants you to go to the station and tell the police he is innocent. He says they will believe you because you know all about the company and the product it is developing ... whatever that means."

"I'm on my way," Rena said after realizing this was no joke.

The sergeant behind the desk at the Medlinton Police Department was courteous but firm. "I'm sorry, Miss, but he can only been seen during visiting hours tonight between six and eight."

"But he is my business partner and I need to talk to him," Rena protested.

"I'm sorry, ma'am. You can come back at six."

"Then I want to bail him out," Rena demanded.

"I'm sorry, ma'am," the officer repeated. "Mr. Ganderson is being held without bond."

"What do you mean? Why can't I pay his bail?" Rena pleaded.

"He is being held as a possible murder suspect and he has to stay in jail until a judge sets a bail amount for him."

"Murder? You can't be serious. He wouldn't hurt a fly."

"I'm sorry, ma'am."

Rena left the police station and drove directly to Silas Berthaby's place of business. After waiting a number of minutes she was shown into the lawyer's office.

Berthaby took his time explaining to Rena that he was a corporate lawyer and his firm did not handle criminal cases. However, he was aware of the circumstances in this case and he would do what he could to help her resolve Brad's problems.

"He hasn't been formally charged yet," Berthaby said. "So far it's only 'suspicion of murder', but it's clear the police think they have enough evidence to charge him with murder or they wouldn't have arrested him. They have seventy-two hours to bring the full charges or they have to let him go. For the seventy-two hours, or until the judge sets bail, he has to stay in jail.

"I think it's best for you to visit Brad tonight and find out his take on what's going on. In the meantime, I'll find out what I can from the police, and we'll put our heads together tomorrow to try to settle on some way to resolve the situation."

That evening, when Rena got to the jail to visit Brad, she was surprised to see Lissy was already there talking to him. Rena had to wait her turn because only one visitor was allowed in the cellblock at a time. Rena identified herself to the jailor, signed her name below Lissy's in the visitor's log book, and settled down to await her turn.

Rena's thoughts were multi-faceted, rapid-fire, and confusing. She resented Lissy taking her and Brad's valuable time. The business they had together had to be more important than whatever Lissy was saying to him, and besides wasn't it clear that Brad loved her more than Lissy because he had expressed it so forcefully on the roof that day. She felt her body spontaneously react to the thought of Brad making love to her and she glanced around the waiting room to see if others had noticed the dreamy look of sensuousness on her face. Subconsciously, she attempted to cloak her feelings by pressing her knees tightly together and squeezing the lapels of her dress together over her chest.

Rena was in this introverted posture when Lissy walked into the room. Lissy never failed to observe the things around her, especially when it had to do with a woman who was attractive enough to be in competition with her. She thought to say something clever, but the formality of the environment, and the gravity of the situation forbade it.

Lissy was not smiling when she said, "Rena, I'm so glad you're here. Brad is in bad shape … he needs all the support he can get."

The two women took each other's hands and stood face to face silently for a moment before Rena spoke.

"What's going on Lissy? Why is Brad in jail? Who told you about it? When …?"

Rena's rapid fire questioning was interrupted when Lissy said, "Hold on Rena. You're asking too many things at once. Let's sit for a minute, and I'll fill you in as best I can."

Lissy and Rena sat together on a nearby bench still holding hands.

"I got the word this morning from Brad's lawyer," Lissy began. "He said Brad asked him to call me. He said Brad could be in some serious trouble and when Brad and I talked a while ago, Brad agreed he may be in over his head."

"Do you think Brad has done something he shouldn't have?" Rena wondered.

"I don't think anything of the sort," Lissy replied with a touch of indignation. "I'm just telling you that Brad could have told the police too much. By that I mean, he told them how he thought the deaths at Gordon's could have happened, and he thinks the cops thought the only way for him to have all that information is for him to have committed the crime."

"So, they arrested him for nothing more than that?" Rena did not understand it. "Excuse me, Lissy, I want to go in and talk to Brad before visiting hours are over. We'll talk more, later." And with that, Rena signaled the jailor to let her pass.

As Lissy had told Rena, Brad was not at his best. It was clear the ordeal of the arrest was wearing on him. Rena had never seen him in such a sullen mood.

"Are you all right," she asked when he sat opposite her behind the glass.

"Yeah, I'm OK. Maybe just feeling a little sorry for myself," was Brad's wooden reply.

"I wouldn't be very happy either; in your circumstances. Tell me, Brad, exactly what happened."

"I think it all happened because I violated one of my own rules. The one that says 'you don't tell someone everything you know'.

"You remember the cop that came to our office? Nettbron's his name. Well, I told him everything I thought I knew about the fire at Gordon's and he linked the fire and the murders with me because he supposed only a guilty person would have all that inside knowledge."

"It has to take more than that to think a person is guilty of something." Rena still did not understand.

"I told Nettbron I thought the crime had been committed by terrorists who had stolen the secrets to the MothBot from Gordon and were trying to cover up their association."

"That could provide a motive for a terrorist, but what do they think your motive would be to do such a thing?" Rena asked.

"I don't know and they haven't said. I guess they don't have to justify locking me up if the charge is only suspicion of murder." Brad winced noticeably when he said 'murder'.

"Did you happen to mention Sam Murphy?" Rena asked.

"No, and that's probably where I made my mistake." Brad became introspective. "I'm only guessing that Murphy is associated with terrorists, and Nettbron could probably read on my face that I was winging it. I suppose after that he thought that everything I said was fabricated."

"I still don't see how the police would have enough evidence to arrest you. There doesn't seem to be a motive, and all the things you told them were directly related to your work."

None of them would know the basis for the arrest until the next day when Berthaby visited Brad. He had demanded to see the evidence justifying an arrest and the police had complied. Brad's story—the police regarded it as a confession—was equally applicable to a terrorist or to Brad. This alone was not enough to warrant any action on the part of the police. However, they had found in Brad's apartment a small drone with a controller to operate it. The police department's forensic lab had carefully examined the drone and found it to be capable of starting fires with the probe it had sticking out its nose. The police knew that Powervert was developing the drone for surveillance work, but this one had been modified as nothing more than a weapon. It became clear to the police that Brad had made this weapon for the purpose of starting fires, and they decided he had tested it at the Gordon residence. In addition, and by far the most damaging evidence, was that remnants from drones were found where the fires had started, and a police officer had seen Brad at the scene of the crime.

According to the police, Brad must have had some reason to get rid of Jason and Gordon, so he murdered them along with Gordon's wife, and set the house on fire to cover up the crime. This conclusion did not answer all the questions about the deaths and the fire, but the police believed the rest would be forthcoming as their investigation continued.

The police did not have an ironclad motive established, but they were sure it was business related since two of the victims were Brad's co-workers. They speculated that he had modified the drone into a weapon to be sold to terrorists, and Jason and Gordon put their lives at risk when they discovered his plot and threatened to expose him.

"This is all ridiculous," Brad complained. "Just ask Rena or Lissy. They both know the whole story about Sam Murphy spying on me and the company, and about his association with terrorists."

"Maybe they do, but I don't," Berthaby said. "I remember a Sam Murphy though. He had some business dealings with my father. I understood he was killed in a boating accident."

"We all thought that," Brad said, "but he must have survived somehow because he resurfaced some time back spying on us."

Berthaby asked Brad to tell him all about it from the beginning. This time, Brad forgot about 'telling everything he knew' as a possible detriment and he told Berthaby everything he could think of from the time he first discovered the microphone under his table at Grubby's to his being arrested and put in jail.

"Is this the same story you told the police?" Berthaby asked.

"It's the same," Brad said, "except I never mentioned the name ... Sam Murphy."

"Is there any way you can prove any of the things you have said?"

"I already told you. Lissy and Rena know the whole story."

"But you're the one who told it to them. Is that correct?"

"Well, yes, but …."

Brad realized how easy it was for the police to think of his story as the same fabrication he told everyone, but as incriminating as it might seem, he had told the truth. He had to find someone who would believe him and help him get out of this mess.

"I swear I'm telling the truth," he pleaded to Berthaby.

"I believe you," was Berthaby's response. "Let me check into the charges further. Maybe I can find some way to get you out of this."

The 'some way out' came easier than Berthaby expected. When he left Brad, he went to Nettbron's office and found him hovering over a pile of papers on his desk.

"Tell me, Nettbron, is it true that you folks take videos and recording of every arrest you make?"

"Well, we try to whenever we can," Nettbron answered, "you 'legal eagles' are always trying to sidestep your client's guilt by using some technical excuse to get them released."

Berthaby ignored the comment. "Did you make a tape of Ganderson's arrest? And if so, can I see it?"

"Yeah, we taped it, but you have to check with the chief to look at it."

Berthaby went directly to the office of the Chief of Police and asked to view the video recording of Brad's arrest. Berthaby was led to a darkroom where he was left alone to watch the arrest video as much as he wanted. After he had seen it through the second time, he returned to Nettbron's office.

"Tell me, are the words in the recording the only ones said?"

"Yeah, not that I'm surprised," Nettbron admitted. "After they are told that the things they say can be used against them, they usually clam up."

"Then nothing was said in the squad car on the way back to the station?"

"Nothing was said. Why?"

"Did you tell Ganderson he had the right to a lawyer?" Berthaby already knew the answer.

"Sure. I told him of his rights when he was arrested."

"Unless some of the tape is missing, you only told him about remaining silent."

Nettbron could not believe what he was hearing. He had rehearsed the arrest spiel dozens of times to avoid this legal pitfall that had freed dozens of criminals including murderers. He hated the fact that judges looked at the right to be informed as paramount—even to the person's innocence or guilt.

The chief was going to be pissed—big time.

Chapter Thirteen: BACK ON THE TRAIL

"The judge said the case was still open and I am still a 'person of interest'. He also said I was not to leave town without the Court's permission." Brad explained to Lissy as she stood near his favorite table at Grubby's.

"I'm just glad you're back," Lissy said. "Whatever made them think you could do such a thing, I'll never understand."

"Well, I hate to admit it, but the story I told Nettbron did sound like a fairytale although it was the truth. The only part he wasn't told was the part about Sam Murphy, and he's the one who has to be as guilty as sin."

"There's no doubt about that," Lissy agreed, "and now it's up to us to track this guy down and make sure he gets what's coming to him."

"Truthfully, there's something else I didn't talk about," Brad admitted. "I modified the MothBot to be a weapon so I could defend myself against Murphy. He has been out to get me for a long time and he almost succeeded when he got me arrested."

"Do you think Murphy did all this just to frame you for something you didn't do?" Lissy was incredulous.

"Yes, that and something else," Brad concurred. "I think besides framing me, there were two other facets to Murphy's crime. One was probably to keep his identity from being revealed, and the other was to demonstrate the effectiveness of the killer drone."

"To show others, maybe terrorists, the deadly aspects of the MothBot," Lissy added.

Brad only nodded his agreement with Lissy's comment before continuing. "And now that he's proven the MothBot can be used to achieve deadly purposes, and failed at framing me, he could try to further prove the capabilities of the MothBot by sending it on a mission designed to kill me directly.

"I think the defenses I've arranged will protect me, but the situation calls for more than that. We need to figure out how to locate Murphy and his terrorist customers, and stop whatever nasty things they are trying to do."

Lissy did not hesitate to offer some advice. "The last thing we should think of is something that would put your life in danger, but now that the police know about the whole thing, do you think that Murphy would be so brazen as to strike at you directly?"

"You're probably right. He wouldn't verify my innocence by killing me outright, but he would try to frame me again, or set up something that would appear accidental."

"What you need to do then," Lissy suggested, "is to vary your routine each day, so he won't be able to set a trap, and to guard your environment wherever you are."

"I'm ready for that," Brad assured her. "My office, my apartment, and my car are all equipped with scanners that can detect the presence of a drone, and that's sure to be the method Murphy will use to attack me."

"And what do you plan to do if you detect one of these drones?" Lissy asked.

"The simple answer is to launch one of my MothBots to counter-attack it. The real answer is it all depends on the circumstances."

"I'm worried, Brad."

"I'm worried, too. Why don't you close up shop and we'll go to my place and drown our worries with some adult beverages?"

Sam Murphy was contemplating a different sort of drowning. If he timed it just right, he could send a Heliorobo to strike Brad with a jolt of electricity as he walked around the pond that was near his apartment. Brad would fall into the water and everyone would think he had had a heart attack and drowned as a result.

Unfortunately for Murphy, the Contact thought otherwise. "You have satisfied us with the capabilities of your Heliorobo, and you have eliminated any links to us by using its destructive power against our enemies. Now, regardless of your passion to rid yourself of this Brad person, any further exposure of the Heliorobo could jeopardize the success of our mission. You will refrain from using it for your own purposes."

Murphy was livid. Who did this pipsqueak think he was—telling him what to do? Maybe he did represent the path to wealth offered by the people he represented, but everything they had he had brought to them and he was not about to let them tell him what he could or could not do.

"I will do as you wish," Murphy lied. "Is there anything else we need to discuss?"

"Yes," responded the Contact. "My people are continuing with the development of the Heliorobo and the work is coming along as expected. However, it is possible that the improvements being made at Powervert are superior to ours. Is there a way to gather information on the progress they are making without the help of your unfortunate Accomplice who lost is life so tragically?"

"The Heliorobo is perfectly suited to handle such tasks," Murphy smiled wryly. *And Mr. Brad Ganderson just might get in the way.* "As you know, I am capable of manipulating the drone in any way I please, and I know the Powervert property very well. I will arrange to have our Heliorobo visit the warehouse and record the activities going on with their MothBot."

Having satisfied the Contact that his intentions were aligned with the Contact's wishes, Murphy took his leave and went directly to his apartment where he prepared a drone for its mission of surveillance.

It would be easy to fly a Heliorobo into the Powervert warehouse undetected anytime the overhead door was open. There, among the roof rafters, it would remain idle—until called upon to record any worthwhile activity of its counterpart—the MothBot.

With Heliorobo #1 ready to be launched, Murphy began his preparations of Heliorobo #2. He bought a stun gun. The one he selected was able to produce an 18 Amp current of 300 kilovolts. This was enough electricity to incapacitate a 250-pound man for several seconds—enough that if the man fell into a body of water, drowning was a certainty.

The stun gun's circuitry was fitted to one of the Laziguns in a way that it could fire its deadly charge into the rectenna of a hovering Heliorobo. The Heliorobo would then transfer the charge to its own Lazigun and redirect it to any target chosen by the operator. Murphy was not planning on taking any chances when he fired the charge from the stun gun into his archenemy, Mr. Brad Ganderson.

Once Murphy had completed his preparations, he launched H1 on a mission to the Powervert building. Getting it inside would be easy with the overhead door to the warehouse open, but the video console on the control module showed it was closed. The batteries aboard the Heliorobo allowed it to operate for several hours, so Murphy sent the drone around the building looking for another entrance.

As it passed near Brad's office, the scanner on his credenza locked onto the radio signals it was sending and receiving. Brad, busy at his work, almost missed the signal. When he realized the scanner had detected something, he whirled around in his chair to see what it was. The compass needle had just come to a stop and the digital frequency read-out was fading away. Whatever it was was now out of range.

Brad hurried from his office to the laboratory. Before entering, however, he slowed his pace to a casual stroll and entered the room with an air of nonchalance. The two scientists working inside paid little attention to his presence. He was a regular and frequent visitor to their workplace.

"How're things going?" Brad asked.

"Same ol', same ol'," Slade answered.

"Have you guys got any MothBots aloft?" Brad wondered.

"No, why do you ask?"

"I just had this weird feeling that something was flying around my head," Brad responded.

"See what I told you?" Bratley said to Slade. "The guy's gone bonkers."

"You would too, if you had to work with a couple of geeks right out of Roswell," Brad countered as he quickly left the room having gotten in the last word.

Concerned that he might have detected one of Murphy's drones, Brad began to tremble at the thought. He tried his best to alleviate his fears by recalling the sounds the scanner had made. They were radio signals to be sure, but there were dozens of these kinds of signals that the scanner was capable of detecting.

He had to learn how to recognize the unique frequency detected when one of these deadly drones was near. He had to be more attentive next time or the consequences could be disastrous.

A long wait was unnecessary. As soon as Brad returned to his office, he saw the scanner had detected another radio signal. This could be a drone passing outside his office once again. Sure enough, the directional needle pointed to the left and was moving slowly toward the right. Brad pushed the button on the scanner that locked it onto the radio frequency.

If this was indeed the drone, Brad had its number. He set the instrument into its scanning mode again to search for a secondary signal. He knew that the drone would receive its maneuvering instructions on one radio frequency and send whatever video it was taking along with its location on another frequency. So far, Brad had only one of these frequencies locked into the scanner, but he did not know which one it was. Before he could take any counter measures, he had to know the radio frequency of the second signal.

The scanner was not helping. It was detecting one frequency after another; all with equal signal strength. But maybe this was enough of a clue. The stronger signal, the one he had locked on, was probably the drone sending its location and video back to the base station. Brad punched the frequency into his own MothBot controller and set it to 'receive'.

He was in luck. The video monitor on the controller began flashing the view the alien drone was broadcasting. It was the outside of Powervert's building and the drone was circling around toward the back.

Brad was frozen motionless as he watched the video of his own building as this drone continued circling round and round. And then the view changed. The drone was moving away from the building and heading toward Medlinton. Then the video went dark. The drone was clearly on its way to its base and it would follow a pre-programmed route without sending any further signals.

The Heliorobo had been called home reluctantly when Murphy realized the overhead door into Powervert's warehouse would remain closed. Murphy was not worried about his ability to do the things he told his Contact he would do, but he was anxious to get on with it. Tomorrow he would try again when the Heliorobo would be prepared with a fresh set of batteries and recharged with all the hydrogen it could hold. Battery life and light weight would be required for the extended stay at Powervert Murphy was planning. This next time, he would not circle the building waiting for an opportunity. He would command the drone to follow the lunch truck they called the 'roach coach' because he knew the overhead door would be opened as soon as the people inside the building heard the loud horn that signaled 'coffee-break' time.

The next morning, Brad heard the roach coach horn too. He was about to leave his office to join the others outside the warehouse door when he saw the scanner had been activated. The alien drone was back.

Brad quickly set his MothBot controller to receive and record the drone's signals. He watched the monitor as the drone flew inside the warehouse sending video signals of the interior. It seemed to circle around inside looking for something.

The something turned out to be an electrical junction box situated among the ceiling rafters providing a landing place that could be used as a viewing platform.

The drone approached the junction box, circled into position, and landed with its camera showing almost the entire space inside the warehouse.

Brad found it difficult to believe what he was seeing. Obviously, Murphy was adept at maneuvering the drone to manage its operations so skillfully. Brad continued to watch the video monitor as it displayed the empty space inside Powervert's building until the signal faded to black.

A glance at the scanner told him it was no longer receiving radio signals. The drone had come to roost inside the building.

Brad did not know what Murphy was up to, but he knew that with one of his drones inside the warehouse, all he had to do was close the overhead door and the drone would be captured. He wanted to rush out to the warehouse and close the overhead door, but he was aware Murphy might detect his presence and take advantage of the opportunity to command the drone to do its deadly work. Instead, he simply waited until the break time was over and Bratley and Slade had closed the door before he would venture into the warehouse.

No, he could not expose himself that way. Murphy could see Brad's every move if the drone had remained perched where it was. Brad had to think about what was going on and devise some sort of defensive tactic.

It seemed clear the reason for the drone's presence was to set on its perch until Brad entered the warehouse and then send a deadly charge of electricity into him. He knew that when a person died that way, it could be construed as a heart attack and no criminal investigation would be undertaken. But how would that be possible? The drone could not carry enough battery power to provide enough electricity to do the job. *Or could it?*

Brad's thoughts were interrupted when the monitor again showed the view of the inside of the warehouse. After only ten seconds, it went dark again. *What was going on?*

Of course, the drone was being activated periodically to conserve battery power. But what was it looking for if not its chosen victim? Brad was not about to go into the warehouse to find out. He had a better idea. He would spy on the spy craft!

Brad waited impatiently for the workday to end and for Powervert's employees to leave the building. During the wait, his assumption about saving battery power was verified as he observed the video monitor on the controller display its picture for ten seconds every ten minutes. The drone's batteries were sure to last a long time at that rate, but the way the drone was programmed to send its video gave Brad just the opportunity he needed to take action of his own.

He prepared one of his MothBots to be sent into the warehouse. When the latest video from the alien drone faded away, he took the MothBot through the laboratory and into the warehouse where he set it on the floor before returning to the controller in his office.

He prepared to put his pilot training and the MothBot's capabilities to the test as he sent the craft aloft in search of its quarry. He was thankful that he already knew the location of the alien drone as he watched the video the MothBot was sending.

There, up above, he located the electrical junction box. A quick glance at his wristwatch told him there were still a few minutes left before the alien drone's camera would again be activated.

When the MothBot reached the elevation of the junction box, Brad caused his craft to hover as he stared in disbelief. It was an alien drone all right. It looked nothing like the MothBot from which it had been copied.

Brad's first thought was to send his MothBot crashing into the alien drone; destroying them both. Instead, he maneuvered his drone closer until he could see enough space behind and beside the alien drone that he could land there behind its camera's view and remain unnoticed.

Unlike the alien drone, using its power intermittently to conserve battery power, Brad's MothBot was receiving a continuous supply of electricity from a Lazigun in the warehouse. Therefore, he was able to monitor the alien drone without interruption.

As long as it remained where it was, the alien drone would not be able to detect the MothBot's presence, and Brad was confident that it would stay there throughout the night since there was no activity in the warehouse to be observed.

On the other hand, Brad could not stay and watch the monitor continuously, but he had to be prepared to relocate the MothBot if the alien drone changed positions. Otherwise it would be able to see the MothBot and Brad's secret spying would be exposed. He set his video monitor to record the images sent by the MothBot's camera and relaxed in his chair for a long night.

It was the noise of someone entering the building that startled Brad into wakefulness. He did not know what time he had drifted off to sleep, but it had to have been in the wee hours of the morning. After a quick check of the video monitor, he rushed into the bathroom to refresh himself and take a quick shave. He did not want Rena to come in a find him in a disheveled condition. And sure enough, it had been Rena who came to work so noisily, and now Brad could hear her in the break room preparing the morning coffee. Soon the smell of brewing coffee filled the air and Brad was anticipating sipping some of it in his half-starved condition. He anticipated savoring even more the decadent taste of the Frangelico Rena was sure to lace the coffee with.

No sooner had Brad reseated himself at his desk than the monitor of the alien drone's camera showed again the image of the inside of Powervert's warehouse only to darken again after ten seconds. Clearly, Murphy was waiting for something to happen inside that was worth observing and the alien drone would continue its intermittent observations until it did.

Rena was right on schedule with her morning ritual. She would bring Brad coffee with Frangelico and a cup for herself. Then she and Brad would discuss their previous day's activities before formulating future activity plans for their company.

Today, in Brad's mind, Rena's angelic beauty approached sainthood when he saw the coffee treat included one of the most beautiful cinnamon rolls he had ever seen. Trying not to look like the famished animal he felt like, Brad consumed each morsel with passion and satisfaction as he washed them down with that magical elixir only Rena knew how to prepare exactly right.

The two business partners sat and chatted about this and that including the company's business before Brad mentioned the alien drone. Rena's incredulous look turned into one of concern when the camera on the alien drone became activated and Brad showed her the pictures on the monitors.

"That thing can't be something that was stolen from us," Rena decided. "That looks nothing like our MothBot!"

"I know. I thought so too when I first saw it," Brad responded. "But the coincidence is too great to ignore. It seems to have the same features and capabilities as our drone and besides, what's it doing in our warehouse?"

"It's obviously spying; that's what," Rena was adamant.

"Whatever the reason for it being here," Brad said, "we mustn't lose sight of the fact that it's only a tool that's being manipulated by someone who is out to get our secrets, number one, and maybe take me out of the picture in the process. That's why I've got to stay out of the warehouse as long as that drone is in there and it's going to stay in there until its batteries run low."

"I'm worried for you, Brad," Rena said sincerely. "Is there some way we can capture the thing and eliminate the threat?"

"Sure, we could set a trap for it and snare it without too much trouble, but if it's built the way I'm sure it is, it will simply self-destruct once it's in our grasp. Our best bet is to wait until its batteries run low and then follow it to wherever it came from. We're sure to find Murphy at the end of the line and we can put a stop to all this."

"I think that's a great idea," Rena agreed, "and maybe we should help the batteries to run down by giving the drone something interesting to look at in the warehouse."

"Good thinking, Rena. The more its camera is working, the sooner the batteries will run low and it will be forced to return to its base."

"How long do you think it will take?" Rena asked.

"I'd say less than four hours with its camera running continuously," Brad answered.

"Good. We can start right away," Rena suggested. "I'll send Bratley and Slade out into the warehouse with two MothBots. They can fly them in close proximity to one another and that should provide an interesting show for our spy. The boys will only be playing, but the spy drone will have no way to know that. In the meantime, you can set up a MothBot and a Lazigun to supply power to it on the roof above the warehouse door. When the spy drone flies out, you can follow it wherever it goes."

Brad agreed that Rena's plan was workable and they both set out to implement it. Brad was a little apprehensive about being exposed up on the roof, but the equipment was soon set up and he returned to the relative safety of his office. He sent the MothBot on the roof on a check ride and landed it in position above the overhead door to await further action. He knew he would be tested to the extreme to manipulate two MothBots while monitoring three sources of video, but he believed he was up to the task.

The next time the alien drone's camera cycled on, Brad could see the scientists in the warehouse flying their MothBots just as Rena had instructed. Her plan was working. The alien drone's camera did not cycle off after ten seconds the way it usually did. Murphy had to be watching the activity from wherever he was!

The adrenaline of excitement was coursing through Brad's body as he watched the alien drone videoing the scene below. Soon its batteries would be exhausted and then the pursuit would begin.

Yes, even as Brad was thinking about it, the drone's camera faded to black and then re-started almost immediately.

The alien drone was leaving!

Brad watched the video from the MothBot on the junction box as the alien drone flew from its landing place and spiraled downward toward its escape route through the warehouse door.

Ohmygod! Brad thought as he realized they had neglected to open the door. The alien drone had no way to leave the building. It would have to return to its landing place.

Brad moved quickly to fly his MothBot from sight, but it was too late. The camera on the alien drone was showing a video of Brad's MothBot there on the junction box.

The game was over. This drone would not be leading Brad to Murphy.

Trapped as it was, however, the alien drone was not ready to give up easily. It began to maneuver through the complexity of the trussed roof structure. The MothBot was in hot pursuit. Over, under, around, and through the maze of timbers Brad chased the drone. He recognized the fact that with its batteries already low, the drone would not be able to flee much longer, but the thrill of the chase kept him from stopping. This would be a good opportunity to test the MothBot's destructive capabilities and Brad's chance came when the drone neared a corner of the building and would be forced to turn away.

Anticipating the direction of the turn, Brad sent the MothBot in the opposite direction and soon the two drones came to a halt hovering face to face. Brad hesitated only a moment before he sent his MothBot hurtling directly into the alien drone.

Both video monitors went blank.

Slade and Bratley suddenly turned to look when they heard the explosion and saw the huge ball of flame fall to the warehouse floor where it finally became extinguished.

This part of the game was over, but a new phase was about to begin.

Chapter Fourteen: RETALIATION

Sam Murphy was amazed and disgusted at the same time. He was amazed at the agility of the MothBot and Brad's ability to control it even though he was disgusted about the loss of his Heliorobo at the hands of his nemesis. He had watched his video monitor as the MothBot hurtled toward the Heliorobo's camera before everything went black. Not that the drone mattered that much. He was prepared to self-destruct it whenever its presence became known to those it was spying on, but to have it intentionally destroyed by Brad was another matter.

This was entirely the Contact's fault. If he had not insisted that Murphy leave Brad alone and concentrate on gathering information directly from the activities at Powervert, this would not have happened.

Whatever is done, is done, Murphy decided as he contemplated the new tack he would have to take now that Brad had uncovered his drone's stealth operations. Whichever way Murphy's new direction would go, the elimination of Brad Ganderson would be his top priority.

Murphy would send another Heliorobo to monitor Brad's activities with the thought that this would be the way to discover the perfect opportunity to fire a fatal jolt of electricity into his body.

Of the three places Brad was sure to be exposed, one would probably preferential over the others. Going into or out of the Powervert building might be best because Brad did these things regularly at about the same time each day. His visits to Grubby's Tavern were fairly regular and there was likely to be one or more witnesses who would testify that he simply fell over dead from what must have been a heart attack. And arriving or leaving his apartment complex seemed preferential because Brad always walked around a small pond that would be a perfect place for drowning. Any of these places would become the scene of a crime when Murphy's opportunity arose.

Brad had increased his personal security level to 'Code Red'. He knew that Murphy knew he was responsible for the destruction of the alien drone and he believed Murphy would retaliate. What way would be better in Murphy's mind than to fulfill his vow to eliminate Brad?

As comfortable as Brad felt with the scanners he had placed in his home, car, and place of business, as well as the armed MothBots he had standing by, there was still an air of apprehension each time he had to expose himself in an outdoor environment. He felt sure that Murphy was not going to stop pursuing him, so he had to find Murphy first and put an end to his devilishness.

And now, Brad was more apprehensive than ever. He had just left his office where there was no indication an alien drone was near, but when he got into his car the scanner was abuzz with activity.

An alien drone had obviously just arrived on the scene and even though Brad felt safe in his car, he trembled with the thought of how easily his life could end at the hands of Murphy.

Brad believed that this time the alien drone was not on a spying mission; it was out for blood. He had no intention of helping Murphy commit his crime by exposing himself to its lethal potential.

Remembering the MothBot he had placed in the ready position on Powervert's roof, he entered its number into the keypad on his controller and launched it.

He tuned the monitor on the controller to the frequency of the alien drone, and soon the video its camera was taking was visible on the screen. Splitting the screen allowed Brad to also view the video the MothBot was sending as he maneuvered it until both Brad's car and the alien drone were visible. Then he started the car and drove away.

Instead of taking the road toward town, Brad turned in the opposite direction as though he was leaving the city. As expected, the alien drone followed his car for some distance before giving up the pursuit and turning in the opposite direction. Brad immediately pulled over to the side of the road and commanded the MothBot, which had been following the pair at a much higher altitude, to alter its course and follow the alien drone.

It was extremely difficult for Brad to turn his car around, watch the alien drone, and control the MothBot to follow it all at the same time, but he had to do it for fear the two drones would get out of range and he would lose this golden opportunity to locate Murphy.

Fortunately, the commands to the MothBot were not complex because the alien drone flew a straight course at a fixed altitude. It was more of a problem for Brad to turn his car to the street nearest the drone's beeline flight.

It was only a matter of minutes before Brad pulled his car to the curb because he saw the alien drone begin a spiraling turn downward. This was it. He watched the drone's video as it approached the patio of an apartment and begin the landing sequence.

There on the patio stood a man with a controller in his hands. It was Murphy!

Brad was elated that he had been right all along and the lair of his nemesis was now known. His first impulse was to send his MothBot with its destructive power directly into the man's chest. Good judgment prevailed, however, as he directed the MothBot to return to its base at Powervert.

Brad needed a drink. At least he was temporarily free from the threat of the alien drone, so he headed for Grubby's and Lissy, his beautiful companion.

Lissy was not so busy that she could not chat with him as he sat at his favorite table and tried his best to relate the latest happenings to her in a chronological order. When he had finished, he expressed his gratitude that the ordeal would be over as soon as he informed the police and they placed Murphy under arrest.

"Are you sure that's what you want to do?" Lissy asked Brad.

"Of course I'm sure. This guy is trying to kill me and he needs to be behind bars," Brad shot back.

"Well, if you get rid of him, the threat against you will surely be removed, but what about the other guys; the bad guys that are out to terrorize the world by killing innocent people?" Lissy reasoned.

"It's the job of the CIA or the FBI to catch the bad guys. My job is to CYA." Brad was waffling.

"Cover your ass?" Lissy wondered. When Brad nodded, she continued. "Sure, Brad, but you're in a position to do things the Feds can't do and you have equipment they don't have. I think the least you can do is let Murphy lead you to his contacts and do whatever it takes to expose them for what they are."

"Maybe you're right, but it means I would have to drop out of sight until the job got done."

"Now you're talking," Lissy encouraged. "I'll do as much of the leg work as I can. All you have to do is assume some disguise and drive a different car. You can take my car and I'll take yours. According to what you told me,

Murphy last saw you driving out of town. There's no need for you to come back just yet. In the meantime, you can go about your daily routine … as long as Murphy doesn't know it's really you in that bag lady's costume."

"Very funny," Brad was not amused, but the scenario Lissy had outlined made sense. He would devise a way to identify Murphy's accomplices and hopefully put an end to whatever heinous plans they were developing.

"Can we make the car swap at your place tonight?" Brad was anticipating more—much more.

"It might be safer if we trade cars now," Lissy suggested. "We don't know if using the drone is the only way Murphy is watching you. I think you should change the way you look now, as much as you can, and then take my car to my place. Here are the keys. The apartment key is on the ring. I'm closing up at ten and I'll grab some take-out on the way home. Let me have your keys. When you get ready to go, my car is out back and you can use the back door. Just make sure no one follows you."

Wow! I guess I got my orders, Brad thought admiringly as he finished his Coors. He removed his jacket, rolled it up under his arm, and tousled his hair before Lissy let him out through the back door.

Murphy was not happy about recalling his Heliorobo after he determined that Brad was on his way to somewhere the drone could not follow.

Once again, he cursed his Contact for having deviated him from his quest to rid himself of Ganderson. *If not for that idiot,* he thought, *the job would have been done by now.* Murphy prepared for the next meeting by vowing to not let the Contact interfere again.

When he and the Contact met several days later, Murphy's confrontation with Brad was relegated to a situation of little consequence even though Murphy had revealed all that had happened.

The Contact had other things on his mind. "When you read the newspapers the day after tomorrow," he told Murphy excitedly, "you will see that the development phase of the Heliorobo project is over and the implementation phase has begun."

"By 'implementation' you mean the assignment of a task?" Murphy asked.

"Precisely."

"What more can you tell me?" Murphy prodded.

"You are sure to read all about it in the newspaper," was all the Contact would say.

Whatever the 'implementation' was interested Murphy less than the fact that he had not been included. This left him free to do as he pleased. Besides, he never did like the Contact telling him what and what not to do.

After the meeting ended, Murphy went to his apartment and prepared a Heliorobo to monitor Brad's activities. The previous searches over the course of several days had not resulted in locating his target. Neither had

Brad's car been located at his home or his place of work. Murphy concluded he was still out or town, but he would have to come back sometime, so the monitoring would continue.

And the watcher was being watched.

Each time the scanner in Brad's office announced the presence of another alien drone; Brad would switch his controller to its radio frequency and monitor its video.

Of course, Murphy had no knowledge that Brad was seeing the same things he was seeing. And he had no knowledge that Brad had disguised himself as one of Powervert's workers; railroad cap, bib overalls, and all. And Brad was confident that he would not be discovered each day when he parked Lissy's car in the employees' section of the parking lot.

When he made his regular trip to Grubby's after work, he always parked Lissy's car in a different spot several blocks away, and walked to the establishment. He further attempted to avoid detection by wearing a false mustache, dark glasses, and sitting at different places along the bar.

Lissy enjoyed teasing Brad every time they were out of earshot of other patrons. "Hi there, stranger," she would say, "we're having a special on Coors today. Would you like one?"

Brad would answer, "Sure. That sounds good."

"By the way," she would add, "They say I'm the hottest chick in town. Would you like to come over to my place tonight to prove it?"

"I can tell just by looking," he would say, "but if you can cook, too, I'll be there."

Not all Brad's activities were spent enjoying Lissy's company, doing his regular job at work, and watching the drone's activities.

He found time to send a MothBot to Murphy's place to see if he could catch him in the act of doing something, and once he had seen him leaving his apartment.

Brad directed the MothBot to follow him. Murphy got in his car, drove to the railroad station parking lot, and went inside. Brad wondered if Murphy was taking a trip, but at that time, no train could be seen coming or going. The MothBot was directed to hover around at least until a train got there. No train came.

After what seemed like a long time, Murphy re-emerged and went to his car.

This only could have been a meeting, Brad thought just as another person exited the station alone and went to a car in the parking lot.

Could this be Murphy's connection? Brad wondered. He was not sure what action to take as the second man's car drove away. He increased the altitude of the MothBot so both the railroad station and the second man's car were visible. No other person could be seen leaving the station. Brad decided to follow the car.

The second man drove to a dilapidated warehouse on Fillmore Street in a seedy section of Medlinton where he got out and went inside.

Brad decreased the altitude of the MothBot until he was able to see and make a note of the car's license plate number. Then he commanded the MothBot to regain altitude where it was set in a fixed position with the door the man had entered clearly visible.

Now Brad was free to go about his regular business while keeping an eye on the monitor. The remainder of the day was spent that way and when darkness began approaching, he commanded the MothBot to return to its base on Powervert's roof. The second man's car had not moved and he had not been seen again.

Murphy had not forgotten to check the newspaper on Saturday according to the Contacts wishes. It was unnecessary, however, because the incident had been widely covered by the media. In fact, it had received national attention due to the severity of the incident and the number of casualties it caused.

ARSON SUSPECTED IN
DEADLY CAFÉ BLAZE

MEDLINTON – A fire at the Calories Ø Not Café erupted suddenly trapping several patrons inside and resulting in five fatalities. The number of casualties is expected to rise.

The Friday night rush hour crowd had just reached its peak when fire broke out in several places in the restaurant at once.

According to one witness, "The front window just seemed to explode in a ball of fire and I tried to run out the back door, but there was fire in that direction, too. Lucky for me, I was not far from a side door."

A passerby outside the restaurant said she looked inside and fires seemed to be popping up everywhere.

The fire is under investigation for suspected arson.

Murphy could not fail to see the resemblance between this fire and the one he had set at the Gordon residence.

The Contact had made good on his promise to 'implement' what was to be a reign of terror.

Murphy knew it was coming. Since the second and final terrorist attack on the World Trade Center and the elevation of security awareness, the United States had been relatively protected from indiscriminate killing. Now, people who wished to do harm to the United States and Western culture had found an easier method of delivery.

Murphy was surprised that he had been the one who got it all started, but he was pleased at the same time. And even more pleasing was the fact that Brad, who had been arrested in the Gordon case, was now subject to be taken into custody again because the two fires were so similar.

The similarity of this latest incident to the Gordon fire also caught the attention of Detective Nettbron who was quick to link Brad to the blaze.

As soon as the connection was made, Nettbron dispatched the forensics team to the restaurant to look for

MothBot parts. When he was told the same kind of miniature engine parts found at Gordon's were scattered near the places where the restaurant fire erupted, Nettbron decided he had little choice except to re-arrest Brad.

Something nebulous, something he could not put his finger on kept nagging at Nettbron. It was only when he walked into Powervert's office and asked Rena to tell Brad of his presence did it occur to him that Brad could have had a motive to kill Jason and Gordon; they were business associates and maybe enemies, but what reason could he possibly have had for murdering innocent people? And the act probably was not to test the weapon; it had been tested successfully at Gordon's.

No, this was an act of pure and simple terrorism.

Brad was prepared for the worst when Nettbron walked into his office. Instead, Nettbron acted like he only wanted a consultation.

"It looks like your MothBot is at it again," he said. Before Brad could respond he continued, "Yes, it's true. We found those same miniaturized engine parts at the restaurant. What do you have to tell me about it?"

Brad relaxed visibly at this line of questioning.

"I can say with certainty; it was not one of our MothBots. They are sent bare bones from our supplier and the ones that we receive and modify are carefully numbered and controlled by our research staff."

"The evidence at the restaurant and at Gordon's proves the fires were started by one of your drones," Nettbron countered forcefully.

"By a drone, yes; by one of our drones, no," Brad was equally forceful.

"Look, Brad," Nettbron was being the 'good cop' again, "The reason we are sitting here talking like this and you are not in jail calling your lawyer is that it's hard for me to think of you as a terrorist and the restaurant fire was nothing more than that sort of act. But you have to level with me if we're going to get anywhere."

"I told you the truth before and you didn't believe me," Brad countered.

"Go ahead, try me again," Nettbron said sympathetically—almost pleadingly.

"The fires were started by Sam Murphy," Brad blurted out without thinking.

"Well, so now the mysterious terrorist has a name."

"I knew you wouldn't believe me."

"Let's just say I do believe you," Nettbron encouraged. "Tell me more."

Something about Nettbron's deportment made Brad think that he was no longer the enemy, and he began telling him the whole story from the beginning. Nettbron recognized some of the things he was being told as valid because he remembered the incident with Sam Murphy and the boating accident. Brad told him about seeing an alien drone near the Gordon fire and how those same drones were detected nearby, probably with the mission in mind to kill him. He told Nettbron how he had gone undercover fearing for his life.

When Brad had concluded his candid revelation, Nettbron decided to take a chance and team up with him to find and put a stop to Murphy and the rest of the terrorists; whoever they were.

"I guess you're going to have to go to jail after all," Nettbron said pan-faced.

"What?" Brad could not believe his ears.

"I can't think of a better way to keep you safe. After all, we need you and your expertise to help solve this case. We wouldn't want to see you zapped by some alien drone."

Brad was a little relieved, but not completely. "How can I be of any help locked up in the slammer?"

"Don't worry. We'll just take you down to headquarters where you'll be free to roam around as you please. You can also come and go as you please providing you continue to wear a disguise."

Murphy was elated. The newspaper had told the whole story about how Brad Ganderson, a suspect in the Gordon fire, had been arrested in connection with the deadly fire at Calories Ø Not Café.

The authorities had classified both fires as "acts of terrorism" which meant that Brad could be executed if found guilty, and guilt was virtually a certainty since he had been found with incriminating evidence and he had both the motive and the opportunity. Murphy made a note to remind the Contact to refrain from any further overt acts until Brad was sentenced to death.

The following Thursday at 4:00 PM when the two met at the train station, the Contact said he would try to keep any more missions from happening in Medlinton, but stopping the activities altogether would be impossible because his associates had already developed a plan to terrorize the entire United States until its government surrendered, and swore allegiance to the "one and only, true, Almighty God."

As for the present, the destruction of life and property at Calories Ø Not was disappointing. The Contact had considered the fact that this was in many ways a test run, but he expected more non-believers to be casualties. After all, it would be difficult to bring the U.S. to its knees by killing only a handful of people at a time.

Future missions would have to target places where more people were congregated, a crowded disco for example. Several Heliorobos could be sent to firebomb every exit to preclude escape and the number of deaths would surely be in the hundreds.

In addition, the Contact told Murphy of plans to intensify the carnage with the addition of suicide bombers. The successes of this activity in Iraq, Afghanistan, and other places made it easy to recruit men and women wishing to become martyrs.

Some zealots were practically beating down the doors to join this elite group, but allowing them in had been purposely slow because there was always the fear that the government would send an infiltrator and if that person was allowed in, the cause would be lost.

"I don't believe I can contribute to the cause by becoming a martyr, but I am willing to do whatever other duty the movement thinks is of importance," Murphy volunteered.

"It is good to hear you say that, my brother. I will speak with the others and tell you the result the next time we meet."

With that, the Contact got up and walked out of the station.

Chapter Fifteen: CLOSING IN

Brad was almost enjoying being under arrest. He was allowed to roam around the police station as he pleased and his access to resources that would otherwise be impossible for him to use seemed to justify Nettbron's decision to take him into custody.

And Nettbron himself was warming up due to the detailed information Brad was able to provide, along with his willingness to cooperate in the Calories Ø Not fire investigation.

He was happy to provide Brad with information about the owner of the license plate number Brad had videoed at that seedy warehouse. If the person that car was licensed to was indeed one of the members of a terrorist cell, he was destined to be brought to justice, but not before the entire organization could be rounded up and put out of business.

But aside from the business aspects of the new cooperative relationship, Nettbron had found an attraction toward a certain beautiful blonde at Powervert. He used every excuse he could think of to visit Rena as often as possible.

The frequency of the visits did not go unnoticed by Rena. She too was beginning to look forward to the times when he would come and chat with her before going into Brad's office to monitor the scanning equipment.

Nettbron, who had always been abrupt and no-nonsense, was beginning to show signs of warmth and understanding. Rena knew instinctively that much of this was simply the boy/girl thing, but it did not matter. She found him more attractive with each visit and before long the fantasies she often had about Brad faded into oblivion.

During Nettbron's fifth or sixth visit to Powervert, as he spoke with Rena about her role in the organization, he was somewhat surprised to hear she was the managing partner of the company with Brad as the only other general partner.

He was very much surprised to learn how much she knew about Murphy and his heinous activities. Clearly Brad had discussed the matter with her at great length, so he felt they could talk about it freely.

"We already have enough evidence against Mr. Murphy to put him away for a long time," Nettbron said, "but doing so would probably cause his accomplices to be so cautious with their activities we would have little chance catching them."

"I understand," Rena said. "How is the investigation of the others coming along?"

"Well, Brad led us to a guy who meets regularly with Murphy and we are convinced he is one of the cell members. We've been following him since we found out about him and so far, nothing has come of it. Even our wiretaps of his phone have led nowhere."

"Are there any other leads you can follow?" Rena asked.

"No, we keep regular checks on Murphy, but so far this is the only person he contacts."

"I wish there was some way I could help," Rena said. "I would like to see these people out of business as much as anyone."

"There is a way," Nettbron responded. "Put one of Brad's scanners in your car. He has it set on the radio frequency of the alien drone. When the drone is in range, the scanner automatically locks onto the signal and it shows a picture on your monitor of the things the drone is videoing. If this happens and you see a picture of your car on the screen, stay in the car and call me. I think you're safe, but we never know what this crazy idiot is going to do. Another way is for us to keep discussing these things and between the two of us; we're bound to come up with some answers. I suggest we continue our discussion tonight at dinner. What do you think?"

"I can't think of a better plan. Can you pick me up at my place at eight?"

"Sure. Don't forget your scanner."

Murphy was growing impatient waiting for Brad to be convicted and executed. He wondered why he had not thought about it before; the fact that condemned prisoners often spend twenty or more years on 'death row' filing appeal after appeal while waiting their turn. He was not going to wait. He knew he could not get to Brad as long as he was locked in a cell, but prisoners were often moved around to go to court or be transferred to another facility. One of the Heliorobos would be dispatched to the County Courthouse where Brad was being held—to make sure the next time Brad could be seen in public would be his last time.

The Heliorobo Murphy had sent hovered outside one window after another at the courthouse building. The video it was sending allowed Murphy to identify the way the building was designed on the interior. He wanted to locate the courtrooms, the police headquarters, and any other place Brad could possibly be seen.

He knew that the courts took cases beginning at ten and one window presented an excellent view of the waiting area outside the courtrooms. Another window allowed observation of the jail area from which the prisoners would emerge whether they were on their way to court or if they were being transferred someplace else.

Murphy planned to watch the window to the jail area between 8:00 and 9:30 and the courtroom window from 9:30 to 10:00 and again around noon when the court would probably take their lunch break.

The planning, however, proved to be unnecessary because during Murphy's reconnaissance Brad was seen talking to one of the police officers.

If Brad had been seated sweating under a spotlight, Murphy could have understood it, but these two were laughing and joking as though they were old friends. Murphy regarded the scene as highly unusual, but he was in the process of dismissing it when another irregular thing happened. The police officer walked away and Brad took a seat at one of the desks and began shuffling papers as though he was working!

With that, Murphy sensed danger and commanded the Heliorobo to immediately return to its base. He had to give the situation some thought.

With the radio and the TV turned off, Murphy sat in silence and thought about the situation. He had witnessed someone going about his daily business, not someone who was under arrest for murder. No police organization would allow a murder suspect to be in an unsupervised position where escape would not only be possible; it would be easy.

But why was this happening? What possible reason could there be that would justify this scenario?

And then the answer came to Murphy. Brad and the police were getting ready to set a trap for him. Brad must have discovered it was he who had sent the Heliorobo before he destroyed it. But how could that be possible? Even with its near-invisibility, Brad or one of the others must have seen the Heliorobo in Powervert's warehouse, but how could that have led to the person who sent it?

Of course, it was not the warehouse incident that led Brad to Murphy; it must have been the Accomplice, Gordon, who betrayed him by telling Brad everything. And now Brad was working with the police to somehow entrap him. It was something else Murphy could not understand. Why did they want to set a trap when they could simply arrest him? Was it so they could follow him to the Contact? Murphy doubted this reason because he had seen the Contact recently and neither of them had observed anyone following them. And they had always been extremely careful about being followed.

Whatever the reason for their hesitation, Murphy was sure there was enough evidence against him to warrant an arrest. He had been stupid enough to taunt Brad with the Heliorobo at the Gordon fire and Brad had seen the same type drone in Powervert's warehouse. Murphy's relative safety was probably because they did not know where he lived. If the police came to his apartment, they would find the Heliorobos and the controller and Brad could testify they were the same type he had seen previously. Murphy would be caught with a smoking gun; so to speak.

Murphy did not stop to pack. He gathered his shaving equipment and toiletries, his personal effects, the Heliorobos, and the controller before unplugging the telephone from its jack and switching off the main electrical power.

Then he slipped out the back door and drove his car to the long-term parking lot at the airport. From there he took a taxi to the train station. He entered one door of the station and went out another to the hotel across the street where he was sure they would accept cash and not ask any questions.

Murphy stayed at the hotel for two days while he shopped for new clothing and prepared for the next move; the move that would take Brad out of the picture and out of Murphy's life forever.

Rena did as Nettbron had instructed her to do. She made regular visits to the jail under the pretense of seeing Brad. Of course she would see Brad, but not as a visitor sees an incarcerated person. They would sit at Nettbron's desk and talk about business and non-business issues.

The main topic of discussion lately concerned the fact that Murphy had dropped out of sight. They thought their original plan to ferret out the members of the terrorist cell should remain intact because they still had their sights on the person Murphy had led them to. The police had identified the other person as Roland Benjii.

Detectives working under Nettbron's supervision had followed Benjii to the train station hoping to reconnect with Murphy's trail.

Unfortunately, Benjii had only been observed sitting at the station reading a newspaper and talking with no one. Actually, he had spoken to one vagrant who appeared to be asking directions because he held a piece of paper for

Benjii to read. Benjii had looked at the piece and then pointed in a direction. The vagrant immediately left in the direction indicated.

The detectives had failed to observe the 'vagrant' was actually Murphy looking as ragged as he possibly could. The direction paper he had shown Benjii was really a note that said:

> DO NOT SAY ANYTHING
> PRETEND I AM ASKING DIRECTIONS
> I AM STAYING ACROSS THE STREET ROOM 314
> MEET ME THERE AT FIVE – KNOCK TWICE THEN ONCE
> MAKE SURE YOU ARE NOT FOLLOWED
> NOW SAY TWO BLOCKS
> AND POINT TOWARD THE DOOR TO YOUR RIGHT

Murphy's concern deepened as 5:00 o'clock came and went. He was beginning to wonder what his next move would be when he heard a light double-tap on the door followed by a single. He opened the door cautiously to allow Benjii to quickly slip inside.

"I didn't know it until now," Benjii stammered, "but I have been followed by two men. They caused me to be late throwing them off my trail."

"It's all right," Murphy said. "You managed to lose them OK?"

"Yes, but it took some time because I didn't want them to know I was trying to get away from them on purpose. I finally walked into a department store and lost them in the crowd. Clearly both of us are being followed because you used the Heliorobo to attack Gordon and burn down his house." Benjii's accusation did not set well with Murphy.

"I suppose your attack on the Calories Ø Not Café had nothing to do with it," Murphy countered.

"The police might have linked the two strikes together, but it's only because you went against our wishes and used the Heliorobos we were kind enough to give you for your own purposes."

"You seem to forget the Heliorobos came originally from me," Murphy corrected, "but whoever is at fault, we must stop this bickering and work together to resolve the situation. As you can see, I have left my place without leaving any clues as to my whereabouts. And since I already told you I want to become more involved with your operations, this is the perfect time to make the move."

"Very well," Benjii agreed. "And since I too am under surveillance, the operations at the warehouse on Fillmore Street will cease and we will relocate to the country place just outside town. I think you will like it. It's been in existence for some time now. We use it as a training facility."

"Tell me more about it," Murphy entreated.

"Well, it's a fenced-in compound on eighty acres of property that everyone thinks is a religion-based retreat. There are several buildings that are used for instruction and training as well as dormitories. And the acreage includes a large wooded section where field exercises are held. As I said, it has been in existence for some time but only now are we getting the number of recruits we need to achieve our goals.

"Access to the property is controlled with a guard gate, and each person there has been trained to act as though they are members of a religious cult who disdain talking to strangers. They are taught that all people other than themselves are to be treated as outsiders. Our members even dress in pioneer clothing to reinforce their image as devout, religious zealots. You will be welcomed and invited to stay when you come. Please bring appropriate clothing."

With that, Roland Benjii jotted the compound address on a scrap of paper, excused himself, and left the hotel room.

Murphy was pleased with the outcome of their meeting. At last, he would no longer be excluded from others of a like mind. He had been invited into the inner sanctum of the group.

There was one obstacle, however. Living in the compound out of town would restrict his activities in such a way that he would be unable to exact his revenge on Brad. He decided to do the deed while he was still in town and near his objective.

The following morning, he recovered the stun gun enabled Heliorobo from its box inside his rollaboard suitcase, set up the controller, and launched the drone out through the hotel window.

Once again, Rena was in the process of fulfilling her duty to visit Brad at the Medlinton jail. She parked in the 'VISITORS' section of the parking lot and was beginning to get out of her car when a sound from the controller on the passenger seat attracted her attention. She was not sure at first what it was. The controller had never responded this way before. She slipped back into the driver's seat and checked the controller. Its LED readout indicated the numbers of a certain radio frequency and its video monitor was flickering with what was rapidly becoming the view of a window looking into an office.

Rena was aghast. She was watching the video being broadcast by an alien drone, but moreover, she was seeing the inside of the police department with a clear view of Detective Nettbron and BRAD!

Then she watched in amazement as the view changed to another window and then a door and then another door. The drone seemed to be searching for a way inside.

Flustered as to what to do, it took only seconds for Rena to realize that Brad must be notified. Rena punched his cell phone number.

"Brad, you've got to listen to me," she blurted out. "Stay just the way you are and do not look around. I picked up the signal from Murphy's drone. It's outside your building searching for a way in. It's going from door to window to door. Brad if it gets inside, you will be in terrible danger. I'm just outside, but I'm afraid to come in for fear the drone will come in with me. What can I do?"

"Hold on, Rena," was all Brad could think to say. After a few moments of silence he spoke again to Rena. "You must come in, Rena and bring the controller with you."

"But, Brad ...."

"It's all right, Rena. Wait until the drone has moved away from the door you will enter and then come in quickly. I will move around the corner away from any windows, but you will be able to see me when you come in. Do it now, Rena."

Just as Brad finished speaking, the drone's camera was showing a video of the door closest to Rena. Then the view shifted to the side of the building and then a window. Rena grabbed the controller and made a dash for the police station door.

Once inside, she was able to relax somewhat. Only a short while later, she was up the stairs and inside the detective offices. Brad was waiting for her beside the coffee break counter.

"Over here, Rena," he beckoned. She rushed over and placed the controller on the counter, but she could not help trembling.

"It's all right, Rena," Brad consoled. "Now let's take a look at what that drone is seeing."

Nettbron joined them as they watched video from the alien drone circling around and around the building. Brad knew the drone's batteries would not last forever and it soon would be forced to abandon its searching.

Finally the view changed to the building across the street and then to the street itself. The drone was returning to where it had come from.

"Watch closely, guys," Brad warned. "We've got to see where it goes."

Brad and Rena were able to identify some of the businesses along the way by reading their signs. Nettbron, on the other hand, was intimately aware of the way the city was laid out, and he was able to name the street the drone was on and identify which streets it was crossing.

"It's headed for the train station," he surmised. Now Brad recognized the locale, too. He was familiar with the area around the train station because he had followed Benjii from there with the MothBot.

Just as they were thinking the drone would land at or near the station, it took a sharp turn to the left and went directly into a window of the building across the street.

"I know that hotel," Nettbron cried. "What floor was it?"

"Third floor," Rena said, "third window from the corner."

"Nice going," Brad beamed at his partner.

"Terrific," Nettbron joined in. "It looks like we're on the trail of Murphy once again. This time we won't lose him."

Nettbron was quick to follow through. He ordered four detectives to the hotel. They were given photos of Murphy and told that he was not to be lost under any circumstances; nor was he to know that he was being followed.

The intensity in Nettbron's eyes and the gravity of his voice told the detectives that any bungling of this assignment would lead to pounding a beat for the foreseeable future.

Murphy was disappointed with his inability to get to Brad, but he was not ready to give up. Joining the others at Benjii's retreat would have to wait until the task at hand was completed.

In the meantime, he would comply with Benjii's request to assume the role of a devout follower. To accomplish this, he would need the appropriate pioneer garb. He was not sure exactly where to start, but he thought a thrift store might provide the needed clothing. Before leaving the hotel, Murphy put on his vagrant's costume. He was confident that his whereabouts was unknown to the authorities, but he did not want to take a chance that someone might recognize him.

Murphy's disguise did not fool the detectives assigned to him for one moment. Three of them began a leapfrog type surveillance. They would anticipate Murphy's direction of travel and two detectives would move ahead of him to cover two possible routes. The detective who had guessed correctly allowed Murphy to pass and then directed the other two to go ahead to two more possible routes. This way, even if Murphy got a look at one of them, the next one he saw near him would be a different person.

The fourth detective stayed behind at the hotel. He flashed his badge at one of the maids, asked for her passkey, and ordered her to stay inside the janitor's closet. He moved to Room 314, unlocked the door, and secured it in the open position with the night latch before returning to the janitor's closet and giving the maid back her key with a smile and a thanks and a warning not to mention this to anyone. He felt comfortable that she would never know which room he had unlocked and entered.

Inside the room, the detective located Murphy's rollaboard suitcase, found a place to attach a small GPS tracking device, and left the room when he was sure the hallway was clear. Outside, he called Nettbron to make a check of the tracking device and was satisfied to learn the tracker was working properly.

The tracker he had planted was one that remained inert until called upon to emit a signal identifying its GPS location. The sporadic use of this device gave assurance that a searching scanner would not intercept its signal because it was not on long enough to be locked onto.

If Murphy took his suitcase wherever he went, his location could be detected at any time.

Meanwhile, the three detectives followed Murphy to a thrift store where he purchased what could only be described as oddball, used clothing before returning to his hotel.

In all probability this was going to be a long evening and night waiting for something else to happen. Thankfully, there were four of them so they could relieve one another at regular intervals.

Back at police headquarters, Rena, Brad, and Nettbron were not idle.

Nettbron was saying, "I'm sorry to tell you this, Brad, but it looks like Murphy isn't going to give up until he does a number on you."

Tears were beginning to well in Rena's eyes when Brad said, "I know, and I think it's a good idea."

Rena's tears suddenly vanished. "What? Are you crazy?"

"Maybe," Brad responded, "but if Murphy isn't going to stop and we want him to roam free to lead us to the others, we can set it up so he thinks he has killed me. Then hopefully, he will stop bothering with me and take us where we want to go."

"That's preposterous," Nettbron joined the fray. "There is no way putting your life in danger will be tolerated."

"It's possible," Brad retorted, "that we can pull it off without putting my life in danger."

Murphy was not happy about having failed again. His determination to get to Brad, however, caused him to think of an alternative plan. He would send the Heliorobo into the police headquarters with the janitorial crew. Their cleaning duties took them there every night and they often propped doors open to facilitate their work.

That evening under the cover of darkness, Murphy launched again the Heliorobo that had been modified with the deadly stun gun.

Maneuvering down the center of the street high above the streetlights, it was not long before the drone arrived at the police station and was commanded to land on top of the garage in the rear. From this vantage point, the entrance used by the cleaning crew could be easily seen and Murphy waited patiently for their arrival.

As expected, when the crew arrived they propped all the doors open, so their equipment could be moved in easily and their electrical cables would not become entangled by the doors.

This is a piece of cake, Murphy thought as he guided the Heliorobo undetected through the halls and into the room where Brad and Nettbron had been seen.

Panning around the room with the Heliorobo's camera, he spotted a landing place above the kitchen cabinets in the break room. The camera could video most of the main room from here, and the drone was unlikely to be detected.

Murphy shut the Heliorobo down. He would try to get some sleep before 9:00 o'clock the next morning when the building opened for business.

Sleep would not come easily. He could visualize the expression on Brad's face when he noticed the Heliorobo heading directly toward him and realized that his time on earth was at an end.

Murphy awoke with a start. The clock displayed the time as twenty minutes past nine. Murphy was furious with himself for oversleeping. He jumped out of bed, went into the bathroom, and splashed cold water on his face. Not that he really needed it, the adrenaline in his body was already running at a high level, but he wanted to augment his alertness for the task ahead.

He returned to the desk in the hotel room and switched on the controller. The video flashed on with the same view he had seen the night before. Some people could be seen moving about, but there was no sign of Brad or Nettbron.

Murphy's hopes were beginning to fade after he had watched the video for more than a half hour. He decided to make a cup of coffee from the ingredients provided by the hotel, and had just arisen from his chair when first the detective then Brad appeared on the monitor. They walked across the room and out of sight. Murphy's heart was pounding.

He was prepared to launch the Heliorobo as soon as Brad came back into view. This did not happen. Soon Nettbron could be seen again as he seated himself at a desk facing the direction where Brad had disappeared. Nettbron was talking to someone out of the camera's sight.

He must be having a conversation with Brad.

The muscles in Murphy's jaw tightened visibly as he launched the Heliorobo. He guided it out of the break room and into the main room.

No one seemed to notice it at first. Then suddenly, Nettbron saw it and Murphy saw him yelling something in Brad's direction. As the Heliorobo's camera panned that way, Brad could be seen standing in the middle of one of the adjoining rooms.

"DIE, HEATHEN, DIE," Murphy screamed as he sent the Heliorobo hurtling toward Brad. It was within inches of Brad's chest when Murphy fired its stun gun and simultaneously activated the self-destruct sequence.

As Murphy's monitor faded to black, he knew the Heliorobo had done its job. He had finally done to Brad what he had sworn to do so long ago.

Chapter Sixteen: THE COMPOUND

"Sam Murphy" was the only thing said to the gate guard, but it was enough to allow Murphy to enter Benjii's compound.

He chuckled to himself when he saw two women entering one of the buildings dressed in those ridiculous costumes. Since Murphy was a bottom to top observer, he first noticed the black leather high-top lace-up medium-heeled shoes. A little bit of white petticoat could be seen under a full length homemade long sleeved gingham dress topped off with a white prairie-type sun bonnet. He saw no faces, but expected if he did they would be devoid of makeup.

The next person Murphy saw was a man in similar attire. His shoes were like the women's except for the flat heels. His homemade trousers held up with suspenders were too short and cuffless. The long sleeved striped shirt he wore was bigger than it should have been—it could have been a hand-me-down from a larger person—and his oval crowned black hat had a wide brim that curled up all around the edges.

If I have to dress like this, I'm leaving, Murphy thought as he sought directions to Benjii's location.

"Where can I find Roland Benjii?" he asked the first person he encountered.

"I don't know anyone by that name," was the curt answer, "but the main office is in that building." The person pointed to a two-story building nearby.

When Murphy entered the building, he could see that it was a former military barracks that had been remodeled although it still retained its basic characteristics. From the entry vestibule, one could climb the stairs to the second floor; go to the toilet facilities on the right, or the barracks area to the left.

Murphy turned left and came face to face with Benjii just inside the doorway to a large, open, office-type area.

"Welcome, Sam," Benjii greeted him warmly. "Have you come to stay?"

"Yes," Murphy replied gesturing toward his rollaboard suitcase.

"May I ask the means you used to get here?" Benjii queried.

"Of course," Murphy answered. "I made sure I wasn't followed from the hotel by picking the lock to the connecting room and waiting there until I heard someone entering my room. Then I simply walked out of the room and down the stairs to the outside exit. Outside, I walked in the wrong direction for two blocks before doubling back and walking to the edge of town where I hitchhiked a ride to a place two miles past this turnoff. Then I walked back here."

"Very good. Did you bring the appropriate clothing?"

"Mine isn't as homespun as that I have seen so far, but I'm sure it will do."

"Good. Now you can leave your bag here. Martha will watch it for you. You will be staying in this building; upstairs. First though, I want to show you around."

The compound contained two distinct areas. The half nearest the road could have been a retreat. Its several reconstructed barracks buildings attested to the fact that it was made to house people, but its acreage of planted organic vegetables and rows of fruit trees identified it as a place where an Apostolic Socialist subculture could self-sustain their lifestyle.

The entire perception of the property changed where the back half began. The rows of fruit trees transitioned into a dense forest of maple, beech, and dogwood as the ground rose above the bucolic scene below. The forest needed only to be penetrated a few meters before coming to a large open area. Grass from the former meadow remained in only a few places due to the training activity of the area's occupants. And it was bustling with activity.

Close to one hundred people, both men and women could be seen practicing the use of weapons and various forms of hand-to-hand combat. Groups of eight or nine perfected their deadly skills under the direction of a cadre of highly trained individuals.

Benjii told Murphy, "As I said before, we have grown slowly because we searched for only the most dedicated believers. Each person you see has agreed to; even longed to become a martyr for the sake of our cause. Although it is clear that we are willing to end our lives on this earth, this sacrifice is not expected from leaders like you and me because some of us are needed to direct the activities of those pawns toward the achievement of our goals. As you can see, the training is intense even though fighting our enemies is not its purpose."

"I'm afraid I don't understand," Murphy admitted. "Isn't fight training meant to teach people to fight?"

"Normally yes," Benjii agreed, "but in this case, the training is only meant to solidify their commitment to follow instructions. The underlying purpose is to have them obey without question when they are told to strap on bomb vests and detonate them at the targets we have selected."

"And what, may I ask, is the role of 'a leader' like me?" Murphy wondered aloud.

"Your responsibility will be to select several individuals who can be trained to operate our Heliorobos. Only the best people are to be chosen.

"You can use that barn over there for your training exercises. There is a room inside you can use for an office, and you will find the loft contains three dozen controllers and three thousand cartons each containing a Heliorobo.

"Twelve of the controllers are assigned to twenty-four special Heliorobos that are rigged for observation only. The rest of the drones have been armed with explosive gas and incendiary pellets. Their purpose is to destroy the hiding places of our enemies.

"You will call your people 'Devouts' and instruct them on the ways to fly the observation drones to densely populated places where our suicide bombers can be sent to detonate their vests and achieve martyrdom.

Other Devouts will accomplish other forms of martyrdom missions by directing their firebomb laden Heliorobos to places where flames can block the escape of any infidels attempting to flee.

"Now, there are other things I must attend to. You will begin your duties Monday at 7:00 AM. Stay here and observe if you wish. Check out the facilities in the barn. Then go back to the office and Martha will show you to your room."

And with that, Benjii walked away.

Murphy would have found it difficult to leave, even without his instructions to stay. He realized the 'if you wish' part of Benjii's statement was only a formality, but besides his curiosity to see the new training facilities, his reason to linger was caused by a more primordial urging.

It had been a long time since Murphy was around people where he could relax and simply observe. And now that old feeling of arousal was returning as he watched the masculine gyrations of the combatants. The few women present were physically engaging one another as well, but they caused none of these stirrings.

Then one—there would always be one—stood out from the rest. Murphy realized instantly the reason for his attraction to this person. The soft eyes and wide mouth were reminders of Mahinder al Baqr; the first true love of his life. Mahinder had westernized his name to Hinny and adopted the philosophy of the infidels as well. *Good riddance to him*, but the feelings Murphy felt toward him as an object of desire remained.

Suddenly, Murphy realized he had also westernized his name, although he philosophically remained aligned with the 'true believers'. He was back among his people now and the use of a foreign name would no longer be tolerated. He would simply become once again, Sahar.

Martha at the compound's office would be the first to know.

"Please call me Sahar," Murphy instructed her. And Sam Murphy ceased to exist.

I wonder where the name 'Martha' came from, Sahar thought as she escorted him to his quarters.

Sahar's bag had already been placed in his room. It was a simple place, but clean and bright with its window overlooking a pastoral scene below. Yes, over there in a pasture, several head of cattle could be seen grazing their lives away. Sahar would be content here until this phase of his life's mission was completed.

The direction of Sahar's mission was becoming increasingly clear. The confusion caused by his quest to eliminate Brad had suddenly lifted when the Heliorobo's camera revealed it had hit its target directly; ending the existence of both man and machine.

Even though the death of his enemy had been somewhat anticlimactic, it was now possible for Sahar to channel his energies toward new objectives in his newly acquired role of facilitator. And reclaiming his given name along with his long-dormant sexual orientation allowed him to begin the tasks before him with fresh and accelerated vigor.

Monday morning found Sahar in the compound's commissary where he took his morning meal with Benjii and other devotees.

Benjii handed Sahar a list of possible recruits into his cadre of Devouts. Each had shown an aptitude for leadership as well as a zealous nature. Sahar was told he could select from this list or substitute his own candidates. He was not able to link identifying faces with the list of names, but he was hopeful the combatant he had observed before was among those listed. Teaching others to control the Heliorobos was not the only thing on Sahar's mind.

Sahar did not expect to be so fortunate as to select the best Devouts on the first go-around, so he doubled the number of men with the plan in mind to drop the less adept ones until only the cream-of-the-crop remained. Two women were added to the selection of men with the thought that they might be more suited to be controllers than suicide bombers. The women complicated the process because the men wanted to have nothing to do with them, but Sahar was determined to glean out the best people suited for the job regardless of their sex. As it turned out, both women excelled as controllers and were among the selected finalists along with Sahar's dream-boy, Mohammad.

A mental flashback to the way Sahar had developed his relationship with Hinny reminded him of the fact that their affair had never been consummated. He did not want the same thing to happen with Mohammad, but his devotion to duty resulted in the continuous postponement of any overt intimacy.

On and on their platonic relationship continued. And although the training exercises meant daily close contact, Sahar maintained his aloofness even through those times when it was not an easy thing to do.

After only two weeks of training, Sahar was asked to undertake a project with his best three Devout controllers. Benjii explained it was only an extension of the regular training program to identify any flaws a real assault might present.

To carry out the mission, the team was given a white step-van with no identifying markers. Sahar was instructed to drive the van to a nearby town where a concert was to take place and position it on a side street near the concert hall.

One video equipped Heliorobo that had been modified to carry leaflets was to be sent aloft to monitor the activities inside the concert hall. Two other Devouts were to have their Heliorobos standing by to firebomb the exits on Benjii's command. Benjii would direct the action when a crowd had gathered in the lobby during intermission.

The level of adrenalin pumping through a normal person's veins would have been high. This was not the case with Sahar. He had experienced actions like this before and he was confident in the abilities of his team and himself to carry out the mission without a hitch. Sahar's outward calm must have had an effect on the other members in the van as they drove toward their destination because the eyes of each of the individuals revealed no nervousness or anticipation. *Good.* Sahar thought as he studied the map to direct the driver to the place where they would perform their tasks.

By 8:00 o'clock when the concert had begun, Sahar's Devouts were stationed at their controllers and monitors; ready for action. When twenty-five minutes had passed, the rear doors of the van were opened and five Heliorobos flew into the night sky.

Sahar watched with Mohammad, the operator of the video-equipped Heliorobo, as the concert hall came into view. The full height windows of the concert hall revealed a lobby devoid of people. They would wait.

Sahar did not know the whereabouts of Benjii, but he knew his commander could see the action on a monitor tuned to the broadcast frequency of the Heliorobo.

Suddenly Benjii's voice could be heard on the speakers inside the van. "Stand by to send the four fire-bombers inside as soon as the door opens."

The two fire-bomber operators nodded their acquiescence when Sahar looked at them.

At one side of the monitor they were watching, a short, rather fat man could be seen opening the door. He held it open as the four Heliorobos flew inside. The fat man went inside also, but stood near the door. The Heliorobos flew close to the high ceiling; two on each side of the building, and entered the theater area through the mezzanine doorways. Once inside they were made to hover high above the two fire exits in silence.

Inside the van, the operators watched their monitors as the first act ended and the patrons rose to queue up for refreshments in the lobby. From the video being shot from outside, the lobby could be seen filling with people eager to chat and enjoy their favorite drink.

The fat man mingled with the crowd.

"Upon my command," Benjii was saying, "fire-bomb the exits. NOW!" he screamed into his microphone.

An enormous explosion blew out all the windows to the lobby and fire erupted all around. A few people who had survived the blast ran back inside the theater only to see the exit doors burst into flame as the Heliorobos crashed into them. There was no place to turn. Fire was all around and the presence of the dead and the dying forecast a ghastly fate for the others.

"Instruct your Heliorobo to drop its leaflets then bring the drone back here, Mohammed," Sahar commanded. "There is no more to be seen."

The leaflets meant to identify the mission and its organizers fell on the street in front of the concert hall as the Heliorobo returned to the van.

```
THERE IS BUT ONE
        GOD

INFIDELS MUST DIE

    مقدّس حاربـــات
```

Chapter Seventeen: RESURECTION

Chaos had erupted inside police headquarters. Only a few of the officers inside knew what was happening, but everyone heard the explosion when the Heliorobo self-destructed.

Nettbron sprang into action immediately. He ordered every nook and cranny in the building to be searched for "anything that doesn't belong there" as he put it. He did not expect to find another Heliorobo, but an investigation had to be made just to make sure.

With the search underway, Nettbron entered the room where the explosion had taken place and closed the door behind him. There he found Brad face down on the floor. He was unconscious. Hurrying to administer CPR, Nettbron rolled Brad onto his back. Just then Brad's eyes began to flutter as he slowly regained consciousness.

"Where am I?" Brad uttered as he realized Nettbron was kneeling over him.

"Take it easy, Brad. You just got hit with an electrical charge."

"How … what …?" Brad said before he began to remember the circumstances that put him where he was. "Oh yeah, I volunteered to get killed and almost did. I should have realized some of the electricity the drone fired would bounce off that mirror the same way my reflection did."

"You're lucky it wasn't a higher charge," Nettbron said seriously.

"I'll say," Brad agreed before he rolled onto his hands and knees and began searching the floor. "There are bound to be some parts of the drone amongst all this mess."

"Don't worry about it. I'll get our people in here, and I'm sure we'll find the same debris we found at the other places," Nettbron said with confidence, "as soon as the broken glass is cleared away. By the way, that was a stroke of genius to hang that mirror from the ceiling so Murphy would think he was destroying you instead of your reflection."

"Don't think I'm not happy about his mistake also," Brad was smiling again, "but perhaps I should have stood a little farther away; like a mile or two."

The deception that would cause Murphy to think Brad was dead resulted in the necessity for Brad to continue to stay out of sight. His disguise as a prisoner did not work, but when he pretended to be one of Powervert's workers, Murphy had not caught on. Brad would reclaim his overalls and railroad conductor's cap. At least now he would have no fear of being attacked by a deadly flying drone, but he would continue to use his detection equipment just in case.

Although his role as a detective working with Nettbron had been exciting and rewarding, he was anxious to return to his office at Powervert. Nettbron would probably like it too because it would give him more opportunities to be close to Rena.

Things were only beginning to return to normal at police headquarters when Arnie and Sludge approached Nettbron. Arnie did not wear a cap, but if he had, it would have been held twisted in his hands as the stood with his head bowed looking up at his boss.

"You lost him, didn't you?" Nettbron accused.

"Yes, sir."

"May I ask how he managed to get away?" Nettbron was condescending.

"I can't really say, sir," Arnie was fumbling for words. "I ... I don't know how he got out of his room, but when we went in to arrest him, the room was empty."

"So, he disappeared into thin air," Nettbron challenged.

It was Sludge's turn to talk. "We saw him go in the room. He must have left through the window or through the door to the adjoining room, but it was locked."

"So be it," Nettbron sighed. "Tell me, did he take his suitcase with him?"

"Like I said, sir," Arnie responded, "the room was empty when we searched it."

"Very well," Nettbron concluded. "You guys go back to your regular duties. We'll have a meaningful discussion on effective surveillance one day soon."

Lissy could not understand why she had not heard from Brad for so long. She missed seeing him at his favorite table at Grubby's and she wanted him to take her for a late night snack before they both went to her place for the night.

Lissy was a little upset about the situation with Brad, but instead of chastising him about his absence when he called, she expressed her gratitude that he was all right and seemed in good spirits.

Brad told her how much he missed her and he seemed anxious to explain the details of his reason for staying away so long. They agreed to meet at a hotel in a nearby town where it would be unlikely he would be recognized by anyone.

The hotel was equipped for small conventions, so they would be able to take advantage of the amenities and dining facilities the place offered without leaving the premises.

The timing was perfect because Lissy had the following two days off and they would have the opportunity to spend some time together.

The rendezvous began with a passionate bedroom session followed by the relaxation of an hour in the hot tub with bottle of cool Merlot. Lissy wanted Brad to tell his story all along, but she was enjoying the moment too much to interrupt it.

Finally, Brad spoke about his experience. "I'm dead, you know," he said with a smile on his face. "The bad guy finally zapped me with his monster flying drone."

"Stop teasing, Brad," Lissy pouted. "Just tell me what really happened."

Brad told her how his whereabouts had become known to Murphy as he languished his time away as a pretend jailbird. He explained how Rena had discovered the alien drone about to strike him, and how he and Nettbron had rigged the mirror to make Murphy believe he had indeed been successful at killing him. And now, since Murphy had escaped the policemen who were following him, it was necessary to stay out of sight for his own safety until Murphy was apprehended.

Lissy was not ready to hear about Rena working so closely with Brad, but when he explained the budding relationship between Rena and Nettbron, her stance softened. She was even less ready to hear that their normal daily routine would fail to commence again for the foreseeable future. But, in spite of these misgivings, Lissy was determined to deal with the situation as best she could and to be as supportive as possible where her lover was concerned.

The lovebirds were only in the middle of their honeymoon-like rendezvous when Brad's cell phone rang. It was Nettbron with some current news. He had located Murphy—maybe. He could not be exactly sure because the tracking device that had been placed in Murphy's suitcase could lead to someone else if Murphy had transferred ownership of the rollaboard to another person. Anyway, he wanted to know if Brad would be willing to use a MothBot to visit the place the signal was coming from to verify whether the suitcase owner was indeed Murphy.

The veil of sadness that swept across Lissy's face at the thought of their rendezvous ending disappeared when Brad said, "Lissy and I are having fun. If this is something we can do together, I'll do it."

"I'm not sure," Nettbron responded. "The target is located about fifteen miles west of town. You will need to get as close to that location as you can to make your drone work."

A smile flashed across Brad's mouth. He could easily target a location fifteen miles west of Medlinton from where he and Lissy were at that moment.

"I'll do it," he said. "Give me the coordinates and I'll get started."

Brad wasted no time. He already had his controller in the room scanning for alien drones. All he had to do was retrieve a MothBot from the trunk of his car and the surveillance could begin. Lissy snuggled up close to Brad and the monitor. She wanted in on the action. Brad included her by giving her a pad and pen, a highlighter, and a road map of the area. Then he told her to make as many notes as she could about the video they were about to see.

The MothBot flew out the sliding glass door on Brad's command. The distance from their retreat to the coordinates

Nettbron had given took only a few minutes. Before reaching the exact destination, however, Brad caused the drone to overfly the general area; then trace the road back to the main highway. He wanted to make sure they were properly oriented before his final approach.

As Lissy highlighted the roads on the map, Brad directed the MothBot to its target. They were both surprised at the extent of the compound and the fact that the occupants appeared to be members of a religious cult. They were not surprised when the scene changed to show a military style training area. If Murphy was here, it would be difficult to locate him due to the amount of activity and the number of people. The video evidence, on the other hand, indicated Murphy's presence was highly likely. He was just the sort to hang out with fanatics training for combat against the world.

After videoing the compound and the combatants for an hour, Brad summoned the MothBot back to the hotel where its batteries were recharged for the next foray. The recording made from the MothBot's video was downloaded into Brad's laptop where individual scenes could be isolated, magnified, and studied for signs of Murphy. Lissy proved to be adept at the task and shot through the stills much faster than Brad could imagine. No vision of Murphy was to be found. Only one person, seen from the rear, had a physique similar to Murphy's. This person stood out in another way as well. His clothing was old fashioned, like the fifties or sixties maybe, but unlike the pioneer garb worn by the others.

"This could be our guy," Brad observed when Lissy pointed the man out. "Make a note of the time. This could be a regular routine of his and if it is we can get a better look tomorrow at the same time. In fact, tomorrow we should begin videoing just before this scene was captured so we are not looking at exactly the same activity each day."

With the stills examined, Brad touched base with Nettbron who requested that all the videos taken be forwarded to him in case they might be needed for evidence in the future. He seemed pleased at the progress Brad and Lissy were making in such a short time with so little advance notice.

"Nice work, guys," ended their conversation and Brad and Lissy set their spying tools aside and prepared for their last evening at the hotel.

Brad and Lissy ordered room service to bring dinner and wine. They both seemed to have a desire to make this evening a memorable one. The Jacuzzi was filled with hot water and the circulating pump valves were opened wide. The couple was soaking the night away both externally and inwardly. A feeling of euphoria and profound contentment enveloped Lissy as she watched her man relaxing by her side.

"Brad?" she began.

"Um…."

"Do you think I'm a good detective?"

"The best without doubt," Brad replied without opening his eyes.

"I'm serious, Brad," Lissy affirmed.

Brad opened one eye to verify her gravity. She was serious all right.

"What is it, Lissy?"

"Brad, I'm not happy anymore working at Grubby's."

"I understand." Brad's eyes were closed again.

"Brad, I want to quit my job at Grubby's and work with you to resolve the problems you are having trying to stay alive."

Brad opened both eyes and looked directly into hers. What he saw there was compassion and love. She did not want to leave Grubby's because of the job. She wanted to protect him from whatever it was that would threaten him. He wondered if the wine was affecting her; then quickly concluded it was not. Nevertheless, he set out to determine the veracity of her commitment.

"And just how do you propose to help save my life?" he asked.

A lengthy pause preceded her answer. "By doing the same things we did today. To find Murphy and have him arrested. After all, there are lots of things you can't do as long as you're staying out of sight. I can do those things for you."

"OK, and after we rid ourselves of Murphy ... what then?"

"Then I'll find something else to do. I just don't want to be a bartender anymore."

"Very well, as long as you agree that you are but an underling and I am the big boss."

"You're an idiot!" she smiled. "Women have always allowed men to think of themselves as the boss."

The rest of the evening was spent switching dominant physical positions. They both emerged victoriously.

The next morning brought sunshine and smiles to Lissy's new career. She was given her first assignment to deliver the video recording to Nettbron. And at Brad's suggestion, give her resignation notice to the management of Grubby's. If she was allowed to quit and to leave immediately, she was to rent a motor home that had its own electricity generator. The motor home was to be used to pick up Brad at Powervert where they would drive to a camping location near the compound.

"Don't forget to stock the motor home with plenty of food and wine and beer," Brad suggested as Lissy left the hotel.

"And beluga caviar," Lissy quipped as she drove away.

Lissy had been allowed to leave Grubby's employment without the normal waiting period, and when her shopping tasks were successfully accomplished, she drove the motor home to Powervert's office to collect Brad. She almost failed to recognize him in his worker's clothing as he got in the driver's seat and they started for the campground he had located. Once there, they checked in, located an isolated parking space, and then got busy resuming the surveillance activities from the day before.

Lissy and Brad had guessed correctly that a coffee-break time or some other regular activity had taken place when they suspected they might have seen Murphy. Shortly after the MothBot arrived above the compound, the same sort of movement was observed and the same man was seen going from one building to another.

"Let's not get too close," Brad warned. "Murphy is familiar with drones and he could recognize if one is in the area. We'll blow up a still picture when this session is over."

When the duo thought they had enough video showing their target from the front, they sent the MothBot back to the training area once again.

As before, several groups of men and women could be seen engaged in various forms of hand-to-hand combat. Other groups were balancing their way across logs or slithering underneath barbed wire. The scene was full-on military combat training complete with firearms and an observing cadre.

Brad did not know if these scenes would be of value to Nettbron, but he wanted to fill the time he had aloft with a variety of information, just in case. Brad's task of maneuvering the MothBot from place to place was uninterrupted as Lissy busied herself by furiously making notes of the Hollywood movie-style activities below.

After what seemed like a very long time, the MothBot was brought back to its base, and now Brad and Lissy could relax for a moment before trying to identify the people they had videoed.

"That's him!" Lissy cried when the monitor zoomed in on the still photo from the video.

Brad looked closely and agreed there was little doubt about the identity of Sam Murphy. He called and left a message for Nettbron to contact him immediately. Brad wanted Murphy arrested without delay.

To Nettbron, getting credit for a high-profile 'collar' was the name of the game in the detective business. However, he thought that under the circumstance it might be better to spend more time observing the man and the activities at the compound. If they had stumbled upon a cell of terrorists, they were in a position to gather enough evidence to convict the leaders as well as Murphy.

Reluctantly, Brad agreed. Come to think of it, he was enjoying this vacation-like camping with Lissy. Their working as a team was proving fruitful as well as enjoyable. Their mini-vacation, however, was to be cut short after a week had passed.

Nettbron called to advise Brad the tactical unit had gotten permission from a judge to raid the compound and arrest Murphy and his superiors, and to confiscate any terrorist related material. Police helicopters were dispatched to cover the raid and Brad was to stay put and leave his drone in its hangar until further notice.

It turned out to be a long day. The excitement of the chase was over, and the change of pace brought by living in a campground was growing old. Brad and Lissy reminisced about enjoying dinner in a good restaurant and chatting with other people. Along toward evening, they were startled by a knock on the motor home door. They thought it was probably just another camper being friendly, but it turned out to be a forlorn looking Nettbron.

"All right if I come in?" Nettbron asked as Brad and Lissy stood there motionless.

"Yes … yes of course," Brad said as he backed out of the way and motioned the detective in. "What is it? What's wrong?"

"It's just that we blew the whole thing. Can I sit down? Do you have anything to drink?"

Lissy felt like they were failing as hosts as she scrambled to arrange the cushions in a chair and mix a Johnny Walker Black/water/tall for Nettbron. She made regular ones for Brad and herself too. Nothing was said until the drinks were prepared and Lissy had joined the others. Then Nettbron began his story.

"I don't know if the people in the compound got wind of our raid, or if it was nothing more than bad timing, but when the tactical unit got there they found both Murphy and his boss gone. When the SWAT team showed the people Murphy's picture, they just shrugged and said they didn't know who he was; let alone where he was. The guys searched the entire place and they were not to be found. We radioed back to headquarters and asked them to beep the tracking device in Murphy's suitcase just to make sure we weren't crazy and sure enough it beeped back the GPS coordinates and we found the suitcase, but not the suitcase's owner."

"He was there!" Brad exclaimed. "We saw him clearly. Isn't that right, Lissy?"

"That's right," Lissy confirmed, "and you saw him too, Nettbron … on the recordings. Did you guys look around in the training area?"

"Yeah, we looked and we asked the folks there about it. They said it was not a training area at all; only a place where some of the residents came to play paintball."

"Paintball, my ass!" Brad was livid.

"Whether you're upset or not is of no consequence," Nettbron said. "The facts are that we did our best to get these guys but they somehow slipped away. The worst part is now they know we're after them, they're going to be much harder to catch."

"The worst part is," Brad corrected, "I still have to stay undercover. I'm one prisoner who is truly innocent."

"I'm sorry about that, Brad" Nettbron sympathized, "but it's best you stay that way, at least for the time being. One other thing; do either of you think Murphy could have spotted your drone?"

"I don't think so," Lissy responded. "We examined those recordings very carefully and at no time did we see anyone looking back at the camera."

"I noticed that too," Nettbron said, "and I ... hold on a minute."

Nettbron's conversation on his cell phone was one-way; he was listening intently. The expression on his face was one of concern; even horror.

"Turn on the television," he demanded to no one in particular.

Brad grabbed the remote and clicked the set on. There it was on the news; the late breaking story just in. There had been a huge explosion at a concert hall packed with patrons. A fire followed that engulfed the entire building and the loss of lives was expected to be more than two hundred. The cause of the explosion and the fire was yet to be determined, but the authorities have not ruled out an act of terrorism.

Nettbron's dejection was written all over his face.

"I think we've found Murphy," he said.

The speaker crackled in Sahar's van. Obviously, Benjii had something important to say. "Listen ... all of you," he shrieked. "You must abandon the van at once. Our security has been breached. We are all in danger. I am on my way to where you are. When I get there, all the equipment is to be placed in the trunk of my car. Sahar is to get in the car with me. The rest of you are to set fire to the van as soon as we leave. Then you all must leave in different directions. Find shelter the best you can for the night and meet me at the bus station at 11 o'clock tomorrow. Do not check into a hotel. You must not be apprehended. Get ready quickly. I am nearing your position. God is all-powerful."

"You heard what he said," Sahar barked. "Get moving."

Benjii's sedan pulled alongside the van and the trunk popped open. Soon the hardware was placed in the trunk and Sahar had slipped into the passenger's seat. Benjii could see in his rearview mirror the van bursting into flames and the figures around it disappearing into the shadows.

Benjii drove without speaking for some time, but his mood was obvious by the set of his jaw. Sahar wondered what the problem could be. The raid had gone smoothly as far as he knew. He was soon to find out.

"Our compound has been compromised," Benjii began. "It was raided by the police this afternoon."

The tone of Benjii's voice told Sahar that he was regarded as somehow responsible. He said nothing, waiting to hear more details.

"It is good that both of us were gone," Benjii continued. "The police were showing our people a photograph of you. They denied ever seeing you, but the police knew they were lying."

"How could they know that?" Sahar was shaken. He did not want to hear the answer Benjii was about to give.

"They know you were there because they went to your room and the only way they could have known to go to your room is they trailed you somehow."

"But that's impossible," Sahar pleaded. "I told you how I got to the compound. No one could have followed me."

"They didn't have to follow you," Benjii said simply, "they planted a tracking device in your suitcase."

"But, I ...," Sahar searched for words.

"Enough!" Benjii interrupted. "No excuses you can make will justify the way you allowed our operations to be compromised. This means our activities will be forced to go underground, but worst of all, the police will probably link our absence from the compound with the strike we just made at the concert. Circumstantial evidence? Of course, but the police are sure to hound us now until they arrest you, or you somehow disappear."

The color drained from Sahar's face as he anticipated what was coming next.

"Over the next several days, you will groom Mohammad to take over your training responsibilities. When the next opportunity arrives for a mission, you will be assigned a new role."

When Benjii and Sahar reached the compound, Sahar was wary that the police might still be lurking about. Benjii assured him that it would be unlikely that a judge would issue another search warrant since the police had so badly bungled the first one.

He told Sahar it would be safe to return to his barracks, but to take a different room and leave his old room the way it was without disturbing it. Benjii would provide clothing that was more suitable to the retreat environment and Sahar could feel free to go about his business.

Sahar suspected that at least part of the reason he was allowed to return was; if he was captured and arrested, it would take the pressure brought by the authorities off the others.

Chapter Eighteen: ASSUMPTIONS

"We're going to pick up the tab," Nettbron was saying. "The department will pay for the motor home rental and the camp space rental. We really need you two here … keeping an eye on that compound."

"But Rena needs my help at Powervert," Brad complained.

"I'll take care of Rena," Nettbron's innuendo was showing. "Besides, you can communicate with her on your cell phone or by email."

"I suppose that will work for a while," Brad agreed.

Lissy was smiling. She liked the situation just the way it was. As much as she missed relating to other people, this way she had Brad to herself, and it was exciting to use the MothBot to spy on the compound.

The three detectives, one pro and two amateurs settled their plans for the next few days. They assumed that since the attack on the concert hall was over, Murphy would return to the compound. The only thing that might keep him away would be his awareness of the police raid on the place, or one other vague possibility. Murphy could have been the suicide bomber.

At any rate, Nettbron wanted to gather more information about the compound and the MothBot's capabilities provided the best way to do it. Brad and Lissy were to launch their drone at least once each day to record whatever activity seemed to be of interest, and they were to make a few night forays to try to observe Murphy in his room. They were to pay particular attention to whether their drone was detected. If so, they to abort the mission immediately, abandon the campground, and return to their normal activities after notifying Nettbron.

The video Brad and Lissy saw streaming back from the compound was surprising, but it should not have been. The combatants on the practice field were actually shooting paintballs at one another. This activity could hardly be construed as that of a terrorist organization in training.

On the other hand, the people in the refuge section of the compound had not changed noticeably. The principal difference was that Murphy could not be seen moving about as before. The MothBot's nightly trips to his room likewise failed to reveal his presence. Indeed his room remained exactly the way it had been when the tactical team of policeman visited it. Nothing had been disturbed. Nettbron might have guessed correctly that Murphy had sacrificed his life for martyrdom.

The surveillance information when it was passed on to Nettbron led him to presume that further activity along these lines would be a waste of time. Brad and Lissy were told to abandon the project.

"The time has come," Brad said, "for you to decide what your next career is going to be. Our detective work is over."

"And you," Lissy asked, "what will you do, resume your old life now that Murphy is out of the picture?"

"Not just yet," Brad answered. "I want to wait to hear the forensics report from Nettbron. He is going to verify the identity of the suicide bomber when the DNA report comes out. Until then, I'm going to put on my regular working clothes and continue going in and out of Powervert's back door."

"Can I come with you?" Lissy pleaded. "I want to go over the recordings one more time just to make sure we haven't missed anything."

"I can't see any reason why not," Brad concurred. "That will give you and Rena a chance to bond some and give me a chance to catch up on my work."

Normalcy at Powervert had returned—almost. Brad in his worker's clothes and Lissy's presence in his office made things a little different, but when Rena came in with the Frangelico laced coffee, the relaxation of tensions the trio felt permeated the room.

After their break, Brad and Lissy stood around Rena as she watched some of the video shot by the MothBot's camera. Rena was excited about the work they had done and she was disappointed that it had come to an end without capturing Murphy.

Brad and Lissy were competing for Rena's attention as he explained the technical operation of the drone and Lissy talked about the scenes being shown.

It was one of the scenes that caught Rena's attention.

"I know that face," she suddenly said. "No name is coming to me, but I know that face."

Brad and Lissy were curious to see who she was talking about. They both recognized the scene, and it was not one where Murphy was present. It was only a scene of several people in pioneer dress walking about.

"Can we back up the video and run it again?" Rena asked.

Brad began to run the DVD backward until Rena stopped him. Then it was put in the 'PLAY' position again.

"There," Rena cried, "STOP."

The same scene with the same faces as before was frozen on the screen. Brad and Lissy searched each face carefully without recognition.

Then Brad asked, "Which one is it, Rena?"

"That one." She pointed to the screen. She was not pointing to one of the people milling about. She pointed to a face in a nearby window that Brad and Lissy had not noticed before. Recognition was immediate.

"Oh my God," Brad uttered under his breath.

"Who is it? Who is it?" Lissy asked.

"It's Murphy's connection," Brad responded. "It's Roland Benjii!"

"Yes, I remember him now," Rena agreed. "I saw his picture at police headquarters. So what does all this mean?"

"It means we abandoned our surveillance too soon," Brad replied. "The worst of the bad guys is still at the compound."

"We should call Nettbron and tell him," Rena suggested.

"Not just yet," Brad overruled. "We need to be more certain of our identification before we take a chance on making fools of ourselves by pointing out the wrong person."

Lissy was delighted. Now she could go back to her detective work instead of 'finding something meaningful to do'. "When do we start?" she asked.

They all agreed to gather the equipment needed right away and begin the new surveillance the following day. It was Brad's turn to find a motor home at a different company that would be parked in a different spot in a different campground.

Rena went back to her work responsibilities and Lissy hurried away to do the grocery shopping and gather the things they would need for the camping excursion. She arranged to meet Brad at the campsite.

As Brad prepared the MothBots and controllers to be placed back in service, he worried. It could have been an illusion or simply the sight of someone staring off into space, but he was reasonably certain the person they saw in the window was looking directly at the MothBot!

If Benjii had indeed spotted the MothBot, Brad's life would once again be in mortal danger. Not from the hatred of Murphy, but from the protectionism of Benjii.

Brad decided to do some investigation. He returned to the recording and stopped it at the scene with the man's face in the window. Judging by the field of view seen through the camera's lenses, the MothBot had to be about 200 feet away from Benjii. A glance at the geographic coordinates verified the drone to be about 200 feet from the building and about 200 feet above the ground.

Brad took one of his MothBots and set it outside the exit door from his office. Using his controller, he sent the drone flying 200 feet away and 200 feet high. From his window he could see where the drone was, but only barely. Without prior knowledge of its location, Brad doubted anyone would be able to see it with unaided eyes unless something gave away the drone's presence.

Of course, a scanner locked onto the drone's frequency could provide the information just as it had when Brad discovered Murphy's drone. But the person in the window was not looking at a scanner's digital readout. He was only looking out the window. What else could it be?

Brad was beginning to accept the man's gaze as nothing more than coincidental when something caught his eye. He was not sure what caused it, but he was once again staring directly at his MothBot 200 feet away. Brad decided to test his vision again at 250 feet, so he caused the MothBot to move farther away and then he saw the problem.

As the drone began its turn to the new position, the smallest glint of sunlight flashed from somewhere on the airframe. Brad had discovered a fatal flaw in a drone considered to be virtually invisible. A flash of sunlight could have told Benjii the drone was nearby, but could he have recognized it for what it was?

An emergency meeting was called for Powervert's officers and scientists. A correction would have to be made immediately, even if it took all night. No more flat spots that could reflect light were to be found on any MothBot. Brad wanted to be monitoring the compound again the next day and he did not want the drone to advertise its presence the same way bait fish do.

Sahar despised the new items of clothing he had been given to wear. He did not like the barracks-style room he was living in either. He could not understand what went wrong. He had been enduring a level of frustration before, but now his existence seemed to have no purpose. Benjii must have picked up on this when he alluded to Sahar's new duty to blow himself up in order to kill as many people as possible. Sahar wondered how it was possible for him to go so quickly from 'leader' to 'pawn'. Of course, these were designations Benjii was free to make at will. His position gave him the authority to bestow living or dying on whomever he wished. And here he was, the fiercely independent Sahar, at the beck and call of someone else. Sahar decided it was time once again to vacate the place he was in, and start anew someplace else.

It was not going to be so easy this time. All his possessions were locked in his old room leaving him with these rags that would command too much attention outside the gates. The gates themselves were another matter, but first he must find a way to retrieve some of his old clothes. He would wait until nightfall when the others were asleep, then pick the lock to his old room and gather a change of clothes. No one would be the wiser.

During the time between his plan and his departure, Sahar would have to train Mohammad to take over his job. The irony of the situation was not lost on Sahar. He had at last found someone to take the place of his beloved Hinny and now this person was to be taken from him also—and he was helping implement the separation.

The intense hatred he had felt toward Brad had faded away with Brad's death. Now the feeling was welling again, only this time the target for his ire was Roland Benjii. However, unlike the situation with Brad, Sahar's intentions this time were only to get away and distance himself from his newfound enemy.

This decision was practical and not emotional. Sahar knew that the persons sponsoring Benjii and his operations were powerful in their own right. He also knew that if anything happened to Benjii and he was still around, the finger of guilt would point to him immediately and the likelihood of his survival would be nil. Sahar was even concerned about his chances of survival when it was discovered that he had walked away, but he liked these odds much better.

"We're not going to make the same mistake we made last time," Brad was saying. "This time Hector's going with us. It's not fair to leave him home alone with only the neighbor to fill his dish. And besides, he always tells me the right thing to do."

"I'm cool with that," Lissy agreed, "but don't forget … he's your cat and you have to take care of him."

Clearly, Hector was cool with the idea also because he spent most of his time on either Brad's or Lissy's lap, purring at the top of his lungs.

Hector was on Brad's lap, but sleeping soundly when the first MothBot of the new surveillance series was launched. It was nighttime. Brad wanted to use the cover of darkness for the special mission he had planned.

A micro-sized listening device was attached to a servo beneath the fuselage of the MothBot. The plan was to place it near the place Brad had assumed to be the command center of the outdoor training activities. Now it was being used as the place where the paintball guns and 'bullets' were distributed.

It was nothing more really than a poorly maintained shed. The command center occupied a 12 x 12 foot rectangle with four posts that supported the roof. The paintball equipment was stored in a series of lockers along the space where one of the walls should have been. The front space held a counter with an access gate on one side and the other two spaces were open to the elements with only dense weeds and brush growing outside.

Brad maneuvered the MothBot at high altitude until it reached the location of the shed. He then used the video from its onboard camera to direct it to a place where the brush was dense before commanding the servo to release its payload. The listening device fell into the brush and quickly disappeared from sight. Then Brad landed the MothBot atop one of the lockers where it would remain until dawn.

If things went as planned, the light of day would allow Brad to see if the listening device was properly hidden. If it turned out that it was not, he was prepared to fire an electrical charge into it, which would destroy it along with any traces that might lead back to where it came from.

Brad set his alarm for 45 minutes before dawn, munched on the sandwiches Lissy had prepared, and the pair retired for the evening.

Brad was confident the bug he planted was concealed well enough. Hector had kneaded his thigh and purred loudly when the listening device had been dropped.

In the pre-dawn light, Hector's acumen was evident once again. The listening device could not be located by the MothBot's camera. Now the test would be twofold. Would the bug pick up any conversations and would the conversations be meaningful. Brad would have to wait to see.

He caused the MothBot to leave the command center and assume a high altitude where a panoramic view would provide a better chance for observation. The primary mission was to verify the presence of Benjii. If he were truly there, Nettbron would have to be notified so the authorities responsible could capture him and bring him to justice.

The MothBot hovered above the compound for the better part of an hour. A great deal of activity was observed, but no evidence of the presence of Benjii or Murphy was seen. Tomorrow would provide another opportunity, but in the meantime, a more powerful transceiver to pick up the signal from the bug would have to be placed between the compound and Brad's location. The plan was to place it under the cover of darkness also, and tonight was the night.

Dressed in dark clothing from head to toe and guided by the signal from a Magellan Triton 1500 GPS, Brad and Lissy set out to place the transceiver. After only a half-hour following the trail they had created on the Triton by identifying a 'GO TO' point, Brad and Lissy reached the perimeter of the compound. It was easily identified by the six-foot high chain link fence with a coil of razor-wire angled at the top. From where Brad and Lissy stood, the ground sloped downward then rose to a knoll that seemed to be a good place to put the transceiver.

They had only started toward the knoll when Brad stopped in his tracks so abruptly Lissy almost bumped into him from behind. There beside them, partly obscured with overgrowth, was a small wash where water had created a depression beneath the fence. Brad entertained the thought of slipping through the opening and into the compound. Instead, he only pointed it out to Lissy in silence and the pair continued on their way.

The knoll provided an excellent place for the transceiver. It was set on some stones arranged to support it before the entire site was obscured with twigs and branches. They would not know the extent of their success or failure until they got back to the motor home and listened for sounds from the listening device in the compound command center's bushes.

Brad adjusted the Triton's track to 'RETRACE' and the pair set off for the campsite. Neither of them had any interest in any more spy work this night, so they settled for a can of Coors and a glass of Pinot Grigio to wash down their microwave dinner before going to bed.

Morning brought a cloudless day and the smell of one of Brad's famous breakfasts when Lissy aroused from her sleep. Brad was sitting with Hector in his lap and prepared to start without her when she slipped into the dinette beside him. "You're up early," she said. "Have you done any spy work yet?"

"Not yet," he replied, "but I told Hector the things we did last night and he agreed we did the right thing."

"I hope you didn't tell him everything we did," Lissy chided coquettishly.

"No, I told him nothing about that part, but he would have approved if I had," Brad smiled.

After breakfast, Brad tuned his receiver to the frequency of the bugging devices. Noises could be heard in the background, but they could not be identified. Then a murmuring was heard. It became louder until voices could be distinguished.

"Damn," Brad swore aloud.

"What is it?" Lissy asked.

"What a waste. They're speaking in Arabic or some other language I can't understand."

"Let's record it anyway," Lissy suggested. "When Nettbron is included in this, he can have it interpreted."

"Yeah, you're right," Brad concurred as he activated the 'record' function.

No sooner had the act been completed than voices speaking English were heard. Two women were talking.

"Paintball is a game made for children," Voice One said.

"I agree," Voice Two concurred, "but we have to make the best of it until Sahar finishes grooming Mohammad and we can go on with our Heliorobo training."

"Why does Mohammad need to be groomed?" V1 wondered.

"Because Benjii says so, that's why," V2 answered. "I think Mr. B has a special assignment for Sahar, so he wants Mohammad to take his place."

"That's too bad," V1 said with disappointment. "I think he's kind of cute."

"Who, Sahar or Mohammad?" V2 asked.

"Well both, I guess. But I like Sahar's maturity."

"I like it too," V2 agreed, "but I'm afraid you would be wasting your time with him."

"And why would you say that? So you can have him to yourself?"

"Not likely," V2 corrected. "Besides, I think he likes boys better that girls."

"And why would you say that?" V1 repeated.

"Haven't you seen the way he looks at Mohammad?"

"Come to think of it, you're probably right. Do you think that's the reason Mr. B is putting Mohammad in Sahar's place?" V1 asked and then provided her own answer. "To get rid of him because he knows Sahar's a fag and you know how our people feel about homosexuals."

"So why didn't Mr. B send him instead of Saddam to be a martyr at the concert hall?" V2 wondered aloud.

"At that time," V1 guessed, "no one was able to take Sahar's place and teaching the use of the drones to meet our goals is too important. Besides, I think Mr. B wanted to test the resolve and effectiveness of our people and our methods. Now that we know our capabilities, Sahar can be sent on a mission to receive martyrdom."

"Do you know if there are any missions planned?" V2 asked.

"No, but I understand the next one will be the mother of all missions."

Brad was aghast. He could not believe so much information could be gathered so quickly. Technology was astounding.

"Lissy," he yelled. "We just hit the jackpot!"

Technology was indeed astounding in ways Brad had not thought about. He was soon to find out.

Hector was fidgety. Hector was more fidgety that Brad had seen him before. He kept pacing about the space in the motor home, jumping on the couch and then the bed, and meowing in a way that almost sounded like pleading.

"I think your cat wants to go outside," Lissy observed.

Brad was preoccupied. "He can use the litter box in the bathtub."

"Sometimes cats just want to be natural and go outside," Lissy insisted.

"Well, take him out then." Brad did not want to be bothered.

"We already agreed. He's your cat. You take him out. I'm busy fixing us something to eat."

Brad was flustered, but he did not want to fight with Lissy. He stopped what he was doing and gathered up Hector who seemed happy to be listened to. Together, they went out into the dark. Brad put Hector down and he quickly scampered off into the woods. Brad was not worried. Hector would not go far. Brad simply trailed along. After several minutes, Brad began to wonder what the problem was with Hector. He did not seem to want to go to the bathroom. He only continued walking away from the motor home and looking back to make sure Brad was following him.

"That's enough," Brad said after some time. He scooped up Hector and began walking back. Now Hector was squirming. He wanted down. Brad dropped him to the ground where once again he began leading Brad away.

Just then, Brad saw the shadowy figures of two men walking toward his camping place. They must have been fellow campers out for a walk so Brad thought little of it. He was still trying to understand Hector's behavior when there was a bright light from the general direction of the campsite. Brad did not know what it was, but he grabbed Hector anyway and headed toward the glow.

The light from a raging inferno illuminated the same two figures retreating from the place where they had set fire to Brad's motor home. They were carrying gas cans and they were dressed in the pioneer fashion.

Brad dropped Hector and moved up behind the motor home; out of sight of the arsonists. Gasoline had been poured all over the motor home and it was blazing furiously.

Brad grabbed handfuls of dirt and threw them against the back of the vehicle where the storage locker was located. The dirt extinguished the fire on the side of the motor home and on the ground beneath. Brad grabbed the handle to open the storage area door and got a nasty burn for his trouble. He took a handkerchief from his pocket, wrapped it around his hand, and was able to pull the door to its open position. He crawled inside the locker on his hands and knees and once inside, he straightened his legs and raised himself with all his might. The mattress, bedding, and the supporting boards underneath them went flying into the interior of the motor home. Lissy was inside laying flat on the floor. Brad was afraid she had been overcome with smoke, but she was only in that position because she knew the air close to the floor would be the last to become smoke-filled.

Neither of them made an attempt to talk. As soon as Lissy saw Brad, she moved in his direction and together they went back into the storage locker and out into the fresh air.

Many people from the campsite were present using hoses to douse the fire around the main door. None of them saw Brad and Lissy move off into the shadows where they found Hector waiting for them and looking strange as if to say, "I tried to warn you!"

The excitement the camping neighbors were having in their unsuccessful attempt to put out the fire allowed Brad, Lissy, and Hector to leave the scene undetected.

When they got to Lissy's car she unlocked it using a keypad mounted near the door and started the engine with a key she had hidden away for just such an event. They would sequester themselves at Lissy's until new plans could be formulated.

The number one plan on their agenda was to not make the same mistakes again. The attack against them served to strengthen their resolve to bring the hoodlums at the compound to justice. They were now more determined than ever to somehow catch the bad guys red-handed, but this time they planned to err on the side of caution.

Both Brad and Lissy were quick to realize the bug they planted at the compound provided a direct path back to them. Benjii had tried to eliminate whoever it was eavesdropping on his operation by burning them to death along with any information they had gathered. The henchmen he sent to do the job assumed both Brad and Lissy were inside when they doused the area with gasoline and dropped a match into the volatile liquid. And now, as far as Brad and Lissy knew, the terrorists at the compound assumed the two were burned beyond recognition in the motor home fire.

Chapter Nineteen: OBJECTIVE: 'GOTCHA'

The primary adjustment Brad and Lissy made was to observe the activities at the compound from the mobility of Lissy's car. They were not to be sitting ducks again. Besides, this way they could spend their leisure time enjoying the creature comforts at Lissy's apartment. Forays with the MothBot were planned to be made sporadically to avoid the detection of any routine time or place.

Brad believed a link should be maintained with someone who was not a part of the field excursions. Rena was the obvious choice. A code system was established between Rena, Brad, and Lissy. If both Brad or Lissy were all right and things were in order, one of them would either send a text message or call Rena's cell phone and let it ring once whenever they were out in the field. Rena would know who was calling by checking her Caller ID. If there was trouble and help was needed, the person calling would simply let the phone ring until Rena answered. In the event Rena received no calls, she was to call both Brad and Lissy's cell phone to discover the problem. If she got no answer, she was to call Detective Nettbron so he could launch an investigation.

In addition, Brad expanded their monitoring capabilities by teaching Lissy to operate a MothBot. He soon discovered her perspicacity at operating the drone expediently. It was not long before Brad came to the conclusion that this was another area where he would be reluctant to challenge her skills.

Rena was also advancing her ability to operate a MothBot expertly through the tutelage of Bratley and Slade. Before long, the three of them could command a drone to do all the things within its mechanical limitations. Rena underscored her expertise one day by sending her MothBot to observe Brad and Lissy as they, in turn, had their MothBots aloft.

She conference-called them, "Hi guys, look toward the west about 15 degrees above the horizon."

It took a few moments for Brad and Lissy's eyes to focus where they had been told, and then they saw Rena's MothBot hovering very near to them.

"You know my ship has capabilities to shoot down spy drones," Brad said laughing into his cell phone.

"So does mine," Lissy echoed.

"So does mine," Rena retaliated, "but this is only a hugs and kisses mission."

"We love you too, Rena," Brad said. "Now go away. Can't you see we're busy?" And busy they were with their newfound surveillance methods.

Sending MothBots out in tandem brought a new dimension to Brad and Rena's spy operations. Now when one of them observed something interesting, the other could move their drone to see it from another perspective.

As time went by, it was becoming clearer that the MothBots were doing their work undetected. At no time was Brad or Lissy able to see one of the people they were watching; watching them back.

Finally, a break came. Lissy had spotted two people walking toward the command center. She thought one of them could be someone important because he was dressed differently than the other. Brad maneuvered his drone to get a better look. The better-dressed person appeared to be Roland Benjii and the person in pioneer attire looked like Sam Murphy. Brad and Lissy could hardly wait to get back to her place to study the videos in stop-frame in order to get a better look.

At Lissy's that evening, Rena was invited to join in the excitement. Brad connected the video to Lissy's HDTV so each of them would have an equal opportunity for discovery. As the video rolled it became evident that something was in the air. Cult members could be seen hurrying to and fro with a renewed sense of purpose. The eyes of the trio were glued to the TV so as not to miss anything. Even Hector, who usually napped through TV sessions, was watching the screen intently.

When the video finally got to the two people walking, Brad clicked the slow-motion button. He would then cause the video to stop when the men were walking toward the camera at close range.

"Stop it here," demanded Lissy. Brad pushed the 'STOP' button. "Look," she cried, "they are wearing ID badges." Sure enough, identity badges were clearly seen hanging from a chain around the neck of each man.

"Can you blow it up, Brad?" Lissy asked.

"Not here, but the images can be transferred to a computer where magnification is easy."

"Let's do it," Rena joined in. "Maybe we can read the names on those badges."

Brad bookmarked the location they wanted to view better, and then the video was rejected from the DVD player and inserted into Lissy's computer. Brad located the view then centered it on the name badges and began the magnification. Unfortunately, the camera's resolution had not been set to the highest quality. However, as distorted as the names were, the right amount of magnification allowed the letters to be distinguished. The man in the regular clothing was B-E-N-J-I-I and the one in throwback fashion was S-A-H-A-R.

"That's the name those women used at the command center," Lissy observed with excitement.

"That guy is Sam Murphy," Rena contributed.

"There is no doubt," Brad summarized, "Murphy is using the name 'Sahar' and he's walking along with the infamous 'Mr. B'. You can bet those two are up to no good.

It's time we brought Nettbron up to speed. Do you want to do the honors, Rena?"

"Of course I will, I'll call him tomorrow at his office. Maybe he'll buy lunch while I tell him all about it."

"Good," Brad agreed, "and you should take a copy of this recording along with you."

"Rena," Lissy added, "please don't tell him everything you know. We don't want to get into trouble with the law just because we were trying to do the right thing."

"You can count on me," Rena smiled.

Lunch at the Capitol Deli was far from upscale. Nettbron had insisted on meeting Rena there because of work constraints. Nevertheless, they were enjoying one another's company amid the cacophony of the place until Rena mentioned her spy work.

Nettbron was hurt. "I thought you and I were close enough that you could confide in me," he scolded.

"I'm sorry," Rena apologized, "it's just that we had nothing to go on until we spotted Benjii and the guy who now calls himself Sahar. And we noticed something else too. The level of activity at the compound is much greater than before. We think something is up."

Nettbron relaxed. What was done was done and could not be reversed. "Too bad we don't have a listening device in there so we can find out what's going on."

Rena's eyes were focused on the ceiling when Nettbron noticed her nonchalance and said, "Don't worry, I know about the bugging and the fire at the campsite." Rena looked at him with surprise as he continued, "It doesn't take a rocket scientist to figure out what Brad and Lissy were doing when their motor home was burned down by some arsonist. They were just lucky we didn't find their charred remains inside."

"I should have known you'd be on top of it," Rena said admiringly. "So where do we go from here?"

"Well, dinner tonight would be nice," Nettbron suggested.

"Let's have it at my place," Rena agreed.

Nettbron could not have been more pleased.

"Our boy Nettbron is on board," Rena told Brad by way of her cell phone.

"What did you have to do?" Brad asked jokingly.

"Nothing I didn't want to do," was the flippant response.

"I take it that means we can go on with the monitoring uninterrupted," Brad stated.

"All he asks is that you keep me informed, so I can keep him informed," Rena replied. "Actually, I made that last part up as an excuse to be the go-between."

"It's OK with me. You're as much a part of this undertaking as Lissy and I."

"It should be that way. I have as much to lose as anyone if these hoodlums are allowed to attach their terrorist activities to our good name and reputation." Rena was animated.

"I agree, Rena," Brad concurred, "and we're doing our best to put a stop to it."

Brad wondered about how much of a lie he was really telling Rena when he failed to tell her that he and Lissy had decided to take their surveillance up another notch. Nettbron had agreed their monitoring could continue, but he had not defined the word. Brad supposed that could mean monitoring of any kind—like the kind he and Lissy were planning. They intended to sneak into the compound to listen to the plans Benjii and Sahar were making.

Brad and Lissy had discussed the danger of an incursion like this: They remembered how Benjii and Sahar had traced them back to their motor home with deadly efficiency. Once inside the compound, they would be sitting ducks unless their plans worked the way they were intended to work. The first foray would begin this very night.

As soon as the sun went down, Brad and Lissy drove as near as they could to the place where they had found a wash under the compound fence. Brad carried the heaviest part; a car battery and Lissy brought along an electric blanket rolled up like a log and stuffed with newspapers. Their Magellan Triton 1500 GPS led them quickly to the place where they slithered under the fence after first examining it for and any obvious sensing devices. They moved rapidly to a place about two hundred yards inside the compound before placing the electric blanket amongst some bushes and attaching the battery to it. Then they retreated back under the fence and made their way to Lissy's car.

"It will either be a long night or a short one," Brad speculated once they were inside the car. "If they are using thermal energy detectors and some spook inside the compound finds our dummy, the game will be over soon. If we're lucky, they won't know we went inside and we can go on with our plans."

"And you think the battery will last long enough to keep the blanket warm in case they search for us with infra-red?" Lissy asked.

"I believe it will," Brad hoped, "but in the meantime, we better get busy. They could be searching the grounds now."

Lissy had won, or lost, the coin toss to see who sent their MothBot aloft first. Hers was ready to go and she commanded it to fly to the fence and then to the blanket. She would continue this activity for three hours until it became Brad's turn. The observation would alternate this way all night or until their game was discovered. Not much could be seen from what little moonlight there was, but Brad and Lissy were sure that anyone sent to look for them would disclose their presence by using a flashlight. All Lissy could do was pan around the area with the MothBot's camera to see if any light came into view on the monitor.

"By the way, Brad," Lissy remembered, "we didn't wipe out our tracks. If one of those 'spooks' as you call them walks around the perimeter he could see where we went in and out."

"I know," said Brad, "and we would be able to see them. It's better to find out if we're going to be discovered when we're not there." He smiled at his perceived wit. "Get it? They discover we're there when we're not."

"I get it, Brad: Very funny. Now you'd better rest those baby blues. Your turn will be here before you know it."

"A fine girlfriend I have," Brad mumbled to himself as he reclined the back of the passenger's seat and closed his eyes. "She doesn't even know the color of my eyes."

They felt lucky. It was a long night.

Dawn brought a new tactic to their spy work. They left the car and went to the knoll near the breach in the compound's perimeter fence. From there they could see a long stretch of the fence, and it was easier on the eyes to observe things naturally for a while. The MothBots and controllers were brought along in case they were needed for something special.

Brad and Lissy settled in a spot atop the knoll where they were unlikely to be noticed. It was at that time Brad had second thoughts about the trail they had left leading to the dummy.

"Lissy, you stay here and keep an eye on me while I go fetch that dummy and get rid of our tracks. If you see someone coming, text-message me on my cell phone."

"All right," Lissy agreed, "I'll type a warning on my cell phone now, so I'll only have to send it in case of trouble. Be careful, Brad."

Brad took one last look around before moving toward the wash. When he slipped under the fence, he was surprised to see how clear the tracks were he and Lissy had made the night before. He broke a branch from a small shrub and swished it back and forth across the tracks behind him as he made his way to the dummy. Satisfied it had not been disturbed, Brad disconnected the battery. It would remain where it was. Then he prepared to use the dummy to obscure his tracks as he dragged it back to the wash under the fence.

The 'click' he heard was the sound of a safety being disengaged. Brad whirled around to find a Colt 45 aimed at his head. A man dressed in military fatigues growled, "One move and you're dead."

Lissy could not bear to wait for Brad any longer. She sent her MothBot in search of him. When the drone reached the place they had hidden the dummy, Brad was nowhere to be seen. She sent the drone closer. She saw the dummy and the battery some distance away, but Brad was not there. *Something bad must have happened,* she thought as she commanded the MothBot to go to the shed they thought was the command center. Going the entire distance was unnecessary because soon she saw Brad being prodded along with a gun to his back. *Brad has been taken prisoner!*

Lissy wanted desperately to call Rena, but she did not dare take her eyes from the monitor. She directed the MothBot to gain more altitude, and she observed the procession out of the woods, across the training field, and into a barn-like building.

When Brad and the spook disappeared inside, Lissy called Rena. "You have to help me," she cried. "They got Brad and they're holding him prisoner."

"Tell me where you are, Lissy," Rena demanded. "I'm coming right over. I'll call Nettbron on the way."

Brad and his captor entered the barn through the back door. Before them were only empty animal stalls on each side of the open center space. Above the stalls, Brad could see stacks of rows of small cubicles that looked like they might have been salvaged from a shoe store. Something was inside each cubicle, but Brad could not tell what it was. At the far end of the open space where they walked, rooms had been partitioned off; one in each corner with the hallway to the front entrance between. The spook motioned Brad toward the door to their right.

Inside the dimly lighted room, Brad could only barely recognize the man sitting behind his desk. It was Benjii. Brad recognized something else on the credenza behind Benjii's desk. It was a MothBot controller. Brad said nothing.

"We caught this guy snooping around the property," the spook said.

"Very well … put that gun away. You are dismissed." With that the man left the room.

Benjii simply sat there for a moment as though he was sizing Brad up before he said, "You must have seen one of our 'NO TRESPASSING' signs."

"I'm afraid not," Brad lied.

"You had to realize you were entering private property when you managed to get through the fence."

"That thought crossed my mind, but I wasn't sure which side of the fence was private," Brad lied again.

"You should consider yourself fortunate," Benjii advised. "Sometimes the owners of posted private property shoot trespassers and the law says it's OK." With that, Benjii summoned Brad's captor.

"Take this gentleman to the front gate," he ordered. Then to Brad he warned, "See to it that you are not seen around here again."

Brad and his escort had just left Benjii's office when Sahar walked in. He had listened to the conversation between Brad and Benjii, but could hardly believe his ears that Brad had somehow managed to survive the attack at police headquarters. He quickly realized, however, that Brad's very survival had been the mistake he had made that allowed the police to follow him to Benjii's compound. Therefore, he could not reveal to Benjii anything about that situation. He only said, "I heard everything. That was Brad Ganderson. Why did you let him go?"

"Don't be a fool," Benjii rebuked, "you saw the recording the same time as I."

"What does that have to do with it?" Sahar asked.

"The video we made clearly shows two people entering the property. That means the other person knows Brad is inside our compound. That person is sure to notify the authorities if Brad comes up missing or dead and I do not want the cops snooping around here again."

"I suppose you're right. Our infrared detector warned us of trespassers before we sent a Heliorobo to verify it. But why do you suppose they wired that dummy to give off heat?" Sahar wondered.

"The only answer I can think of is they were testing to see if we were using heat sensing equipment. That tells me they were planning an incursion into the compound if they could do so undetected. And that tells me they are looking for some sort of evidence to be used against us."

Benjii's sneer of hatred looked good to Sahar, and the words that followed were music to his ears.

"It is time to bring to an end the work you started so long ago and failed to finish. You and Mohammad must launch your Heliorobos and follow Brad as he leaves the compound. You must not lose sight of him. When he joins the person who is sure to be waiting for him outside somewhere, they are both to be annihilated. Make sure you do not fail again."

Rena was getting ready to call Detective Nettbron when her cell phone rang.

"Rena, it's Lissy. Brad just called. He has been released and he's out of the compound. Maybe you shouldn't call Nettbron if you haven't already done so."

"I haven't called him yet, but I was just getting ready to. Is Brad all right? What did he say?"

"He's OK," Lissy said, "and he's on his way here now."

"All right, Lissy, I'm still coming over to talk to you and Brad. See you soon."

Rena was just approaching the place where Lissy had retreated to her car when she noticed something amiss. Her scanner was indicating an alien drone in the vicinity. She quickly realized that the bad guys were after Brad again, but why were they only following him? *Lissy is in danger!* she thought as she punched Lissy's cell phone number.

"Yes, Rena," Lissy said into her cell phone.

Rena sounded panicky. "Lissy, I'll be driving by your car soon. Listen to me! When I get there, jump in my car as fast as you can. I'll explain later." The phone went dead.

Lissy could hardly bear the next sequence of events. After she got in Rena's car and had heard more about the alien drone following Brad, she realized she had to let him know the danger he was in.

She called his cell phone. "Brad, I know you're near my car because an alien drone is following you and Rena is here and she found it on her scanner and I don't know what to do."

"That's bad news," Brad understated. "You're right. I'm not far from where you parked your car. Are you still there?"

"No, Brad, I'm with Rena in her car, but we're not far away."

"Good," Brad did not want her near any alien drones. "Where's your MothBot?"

"It's still in my car. I left in such a hurry when Rena warned me, I didn't think to bring it."

"If you're talking about your MothBot," Rena said to Lissy, "I have mine right here."

"Rena has hers," Lissy shouted to Brad.

"All right," Brad said. "Now listen. Have Rena lock onto the signal from the drone and note its coordinates. Then you fly her MothBot to a position behind their drone. Hold that position until I get to your car. Once I'm inside, I'll be safe. Then move in behind the drone and crash Rena's MothBot into it. Don't forget to activate the self-destruct function when you make contact with the drone. Is everything clear?"

"I-I think so." Lissy was noticeably shaken, but she was determined to protect her man.

"Just get the MothBot to a position behind the drone. Once I'm in the car, I'll be able to talk you through the process. Hang up now. I'll call Rena's cell phone when I'm safely inside your car." The beep, beep, beep on Lissy's phone told her the connection had been broken.

Lissy's ability to bear the suspense of waiting for Brad to call Rena was somewhat alleviated by the effort she was making trying to carry out Brad's orders. The MothBot had no sooner taken a position behind the alien drone than Rena's ringing cell phone announced that Brad had arrived safely at Lissy's car.

There had never been a better opportunity for the three of them to work as a team and they carried it out to perfection. Brad was getting information from Rena about the alien drone's whereabouts, and then using her to relay instructions to Lissy about how to move in on it and destroy it.

Brad saw the flash of light from the explosive impact when the MothBot slammed into the alien drone and annihilated it.

"We got it!" Brad shouted causing Rena to jerk her cell phone away from her ear. "Congratulations," he said in a more controlled tone.

"That'll teach 'em," Rena speculated as Lissy feigned blowing hot breath onto the fingernails of one hand and polishing them across the fabric of her blouse while she pretended nonchalance.

"Wait a minute," Rena cried, "there is still a signal coming from the same place. How could we have missed?"

They had not missed. The second drone commanded by Mohammad had sent a video of Sahar's drone exploding in flames back to him in the building. He was not sure what had happened. He only knew that Sahar's Heliorobo was no longer available to destroy Brad and Brad's partner, and he was unable to take over the job because Brad was protected inside a car.

The only thing Mohammad could think to do was leave the area—fast. He commanded his Heliorobo to retrace its flight path back to the building.

"Brad, it's not on the same frequency. It's not the same drone," Rena corrected.

Brad said, "Keep tracking it, Rena. Where is it? Which way is it going?"

"It looks like it's going back the same way you came," Rena responded.

"I'm on it," Brad said as he launched Lissy's MothBot out the car window.

Instead of trying to follow the drone back to the compound from which Brad was certain it had come, he set Lissy's MothBot on a course directly to the barn where he had been taken captive. By flying a direct route, he was sure to beat the alien drone there.

Once the barn was reached, Brad instructed the drone to fly around it at altitude. The first time around, Brad saw what he was looking for. The back door was wide open as if waiting for a traveler to come home. Brad flew his craft low, near the ground where he could get a better look at the door opening.

"Where is the alien drone now," he asked Rena.

"I'm not exactly sure, but I think it's somewhere inside the compound," she replied.

Just then, Brad saw it zip inside the barn and disappear from sight. The barn door closed.

Sahar harbored some concern about the way Benjii would react to the failure of putting an end to Brad and his accomplice. His worries were somewhat overshadowed by the fact that Benjii had already condemned him to death by sacrificial suicide and the thought served to strengthen his resolve to escape as fast as he could. The new obsession to remove himself from his surroundings replaced the fervor he once felt for the destruction of Brad. From now on, Mr. Ganderson would be a 'thorn in the side' for Roland Benjii to deal with.

"Sahar, what happened to your Heliorobo?" Mohammad asked as soon as his craft had safely landed.

"I don't know; my screen just went blank, but I do know this; Benjii is not going to be happy about our failure to kill Brad. We should have sent him to his grave as soon as he left instead of waiting to find his accomplice … if he even had someone waiting for him," Sahar snarled.

"You're right," Mohammad agreed. "After your Helio blew up, I watched Brad reach his car and no one else was around. And once he was inside, he was protected, so I brought by Helio back here."

"Whatever the cause," Sahar said, "we missed our target. Benjii must know it is impossible to carry out a mission when the equipment fails, and I think a malfunction in the Heliorobo's self-destruct mechanism caused it to explode unexpectedly."

Sahar knew that they would both be dealing with Benjii's ire soon enough. For now, he wanted to forget about the Heliorobos and concentrate on a plan to escape. Perhaps Mohammad could be persuaded to join him. *Wouldn't that be nice?*

Chapter Twenty: DEADLY MISSION

One of Roland Benjii's Devouts told him about the failure of Sahar and Mohammad to carry out their orders successfully. He was beside himself with fury. He considered beheading them both in the training area to use them as examples of the destiny of those who fail. After some consideration, he decided they would be of greater value when they were sent on missions of martyrdom, and the mother of all missions was before him on the drawing board.

The plans called for three simultaneous raids of three different basketball games in three different cities. Benjii's eyes sparkled with glee at the thought of all those non-believers being trapped inside those burning sports arenas. It would be unfortunate that some of the basketball players who professed faith in Benjii's cause would also die, but they were all rich with the decadence that their profession provided, and their loss would be negligible.

Benjii planned to send six suicide bombers to each location. They would wait outside the building until a commando-style tactical unit neutralized the security forces with grenades and AK-47s. When there was no more effective security and the doors were wide open, the martyrs would rush inside and detonate their vests among the largest group of spectators they could find. While all this was happening, twenty Heliorobos would incinerate all the doorways both inside and outside the building.

Benjii's plan also included a back-up system. An additional ten suicide bombers would be standing by at each site in the event they were needed. If the mission went as planned, the back-up team would not be needed, but they would still have the opportunity to learn and gain experience from their on-site training.

The result of a successful mission was sure to advance Benjii's standing amongst his peers and also qualify him for an elevation in rank to that of a living martyr.

Benjii decided that the training for each specific mission was to begin immediately. He assembled his cadre and carefully outlined the plans. He was cautious, however, to withhold the names of the targets and their locations. He knew that if even one person had this information, a possible leak would be created, and Benjii was not about to take that chance. Failure was not an option.

Brad had a mission in mind too. He was tired of running. He was tired of hiding. He was tired of wearing a disguise and fearing for his life. He would make the switch from defense to offense, and put a stop to all this nonsense.

When he told Lissy he wanted to stop running and start chasing, she was all for it.

"I think it's a good idea, Brad, but how are you going to do it?"

"I'm not sure," he answered, "I think we should call our team together and get some input from everyone."

When the Powervert people were assembled, Brad brought Bratley and Slade up to speed. Many of the details that he and Lissy and Rena knew were omitted, but the points about Brad's persecution were convincing enough to excite the scientists into participating willingly.

Slade was the first to offer, "Going on the offensive is fine, but I hope you're not planning on hurting anyone."

"Honestly," Brad responded, "I wouldn't mind if Sam Murphy got in the way and got hurt, but that isn't the mission I have in mind. I want to figure out a way to neutralize their drones. You know they copied our MothBot and created their own version ... I think it's called a Heliorobo. Anyway, they have modified our surveillance drone into a Heliorobo made to electrocute people and start fires. This is not guesswork because the drone has been used in an attempt to kill me."

"That's true," Rena agreed. "Brad was only able to save himself by fooling the drone operator into thinking he was somewhere else.

"Anyway, the point is that these drones are dangerous when they are in the hands of the wrong people, and the people I'm talking about are nothing more than terrorists and murderers. The police have evidence that the killings and the fires at Gordon's house, the Calories Not Café, and that concert hall out of town all involved the use of these drones."

"Actually," Brad said, "the police think the drones used to do the damage were ours, and that's why I got arrested."

It was Bratley's turn to contribute to the discussion. "Why don't the police arrest all these guys and confiscate their Heliorobos? That should solve the whole problem."

"I know the answer to that one," Lissy beamed; proud of her inside knowledge. "The police raided their compound once and couldn't find the person they were looking for. At the time, they didn't know about the Heliorobos being there and they didn't look for them. Now the Court won't issue another search warrant because the police bungled the last one."

"So, the plan is to take the law into our own hands?" Slade challenged.

"Not at all," Brad objected. "As I said before, all I want to do is stop these guys from using the Heliorobos against innocent people, and I know how it can be done with your help.

"When I was inside the compound, I happened to see where dozens if not hundreds of Heliorobos are stored inside a barn. We could send a couple of MothBots to the location and crash them into the Heliorobos."

"And since they are all filled with hydrogen," Slade observed, "the detonation of one would cause all the others to explode in a chain reaction."

"That's it," Brad concurred. "Our actions will surely cause a fire that will burn down the barn, but that is nothing compared to what these madmen do."

"I'm all for it," Rena voiced, "anyone with me?"

Each team member's hand was raised; one by one. The offensive would begin.

Sahar was unhappy with the pioneer garb he was forced to wear. He was unhappier still with the bomb-laden vest he was issued and told to put on. He was careful, however, not to let his attitude show because, after all, those seeking martyrdom were expected to do so enthusiastically.

The orders had come straight from Benjii. Mohammad was to take Sahar's place and Sahar was to join the suicide squad.

Sahar had made overtures toward Mohammad in order to have a partner to help with his escape. It was not to be. The spark of zeal was still shining brightly in Mohammad's eyes, and he would not be diverted from his cause. Sahar would have to find a way out by himself.

Sahar's assembled comrades consisted of forty-eight men and no women. Each person had been assigned to one of three groups and given a task. Each group would have six bombers and ten standbys. The eighteen bombers would rush their respective targets and detonate their vests. The remaining thirty would stand by in case back up was needed.

All the fanatics voluntarily vied for a spot in one of the bomber groups—everyone that is, except Sahar. He was ordered to be bomber number one, in group number one. Martyrdom was near.

Anticipation, intensified by time constraints, permeated the groups. Sahar knew the moment for action was near. He had to think of something fast. It crossed his mind to go along with the raid and make his escape during the actual assault. This thought was quickly dismissed, however, when he discovered the bomb vests they wore could be remotely detonated by someone other than the bomber himself. He thought to somehow disarm the blast mechanism, but he had no access to the vests other than during the training exercises. Time was running out.

A glimmer of hope that Sahar might be able to get away came from an unexpected source.

Mohammad interrupted the training session under way to tell Sahar his expertise was needed immediately. The compound's security system had discovered a series of strange radio signals that could not be identified. The signals were on the frequencies normally used to control the Heliorobos, but the ones flying in the current training session were performing as expected. Mohammad asked Sahar to help identify the source.

"It's really weird," he said. "The signals security is picking up are nothing but gibberish."

Seizing the moment, Sahar excused himself from the group and followed Mohammad to the building where the Heliorobo training was under way. In his haste, Sahar neglected to remove the deadly vest that remained hidden beneath his clothing.

The plans Brad's Powervert team had made were simple. The five of them would get into Bratley's van with their equipment, go to a strategically advantageous position near the compound, and launch their assault on the Heliorobo building from there. A generator in the van would supply all the electrical power they would need.

All had proceeded as planned until they launched their MothBots and discovered the building provided no access because all the doors and windows were closed. Nevertheless, they continued to command the MothBots to circle the building until an opening became available.

As serious as their mission was, the team found it enjoyable because they were flying their drones the way a gamer would enjoy a game. They had each put on a pair of the special goggles that Derek had introduced them to. This way, they could watch the action through small TV screens built into the lenses. Now the experience was not unlike actually being in the cockpit of the MothBot. In addition, the terrorists had never detected their presence, so they felt free to maneuver about as they pleased. This was soon to change.

When Sahar and Mohammad reached the Heliorobo building, Sahar resumed command by seating himself at his old desk. The Heliorobo monitors he saw before him were displaying something he had never seen before. He instantly knew what it was, but he allowed Mohammad to speak first.

"You see, the monitors are tuned to the unidentified frequencies and they are showing bits and pieces of video," Mohammad explained. "It's like they are encoded somehow."

Sahar responded, "That's because the signals are scrambled and that can only mean one thing. MothBots are maneuvering in our vicinity."

"I don't understand," Mohammad protested.

"Our enemies have copied our Heliorobo technology and they are in the process of using it against us," Sahar told him. "The only difference is they have encrypted their radio signals so we can't see the things their cameras are seeing. The only way we can protect ourselves from the invaders is to annihilate them with our own superior Heliorobos."

"How can we do that?" Mohammad wondered. "We can't even make any sense of their radio communication equipment."

"We don't have to," Sahar assured him. "We will simply send our drones aloft and locate each of the MothBots with our directional sensing equipment. Then all we have to do is follow the indicator needle to the enemy and destroy him with a bolt of electricity. Shall we begin?"

It was Rena who first sounded the alarm. "BRAD," she screamed, "my monitor just picked up some Heliorobos, a lot of them!"

The Powervert team members were beginning to check their own monitors when Rena shouted again.

"One of them has a MothBot in its sights ...."
Slade cut her declaration short. "My TV just blacked out," he said.

"They got you," Brad declared. "Take over the monitors while the rest of us make sure we don't get shot down."

The scene in the sky above the Heliorobo building displayed a flurry of activity. MothBots and Heliorobos took turns pursuing and being pursued. An observer on the ground would not be able to see or hear the chaotic dog-fighting going on above if it were not for the brilliant flashes from one of the drones exploding in flames as it was hit by an alien jolt of electricity.

Brad had just fired at a Heliorobo and watched the scene in his goggles as the drone fell to the earth in a ball of flames when his own MothBot was attacked from the rear and his monitor went blank. He quickly launched another and soon was back in the fray of things.

As the fighting continued it was becoming obvious that the superior number of Heliorobos would eventually overwhelm the MothBots. Brad was not ready to accept defeat.

"All right, gang," he shouted, "let's show 'em how it's done! Fly your MothBots backwards at top speed. That way, you can see the enemy as he approaches, and you can get him before he gets you."

Immediately, the superior level of training of the Powervert team became evident. They expertly maneuvered their MothBots in reverse, and fired their Laziguns each time a Heliorobo came into view.

The generator in Bratley's van provided enough power to send electricity farther than the Heliorobos could manage with their battery power alone, and the enemy drones— falling in flames—were lighting the sky.

Skillful flying by the Powervert team could be given credit for the success they were having, but at least part of the reason was that the Heliorobo operators could not tell whether the MothBots were flying backwards or forward— until it was too late.

Sahar's recognition of the hopelessness of their situation coincided with the appearance of Benjii. Before he could explain to Benjii the danger they were all in, Benjii interrupted him.

"SAHAR," he screamed, "what are you doing here?"

Sahar thought to tell him, and then decided against it. He offered no explanation.

"This is Mohammad's responsibility now," Benjii was still screaming. "Get out and go back to your training. The climax to our mission is drawing near."

He doesn't know just how near it is, Sahar said to himself. Without speaking, he arose and somewhat timidly left the building.

Benjii turned to Mohammad. "What is going on?" he asked.

Mohammad tried to contain himself and overcompensated by saying, "It's just that we are under attack by drones just like ours."

"What?"

"Yes, look at the monitors. Each time you see a screen go black, one of our Heliorobos has been lost."

Benjii watched a monitor in disbelief only to see the screen fade to black. "Idiot," he yelled to Mohammad. "Command your drones to return before they are all lost."

Hope disappeared from Mohammad's face as he told his Devouts to withdraw.

Brad noticed the retreat of the Heliorobos immediately. "After them," he told his team.

Audible sounds of relief could be heard throughout the interior of the van when the team at last could escape the intensity of flying their drones backward and begin to pursue the enemy while moving forward.

The new direction produced even more kills than the previous one. Even though the enemy was fleeing as fast as possible, the MothBot's superior firing range knocked out one Heliorobo after another.

Soon the last of the Heliorobos was seen dashing inside the building that was their base.

"'Inside! After them!" Brad told his team as he saw the building's doors closing rapidly. "Now stop and hover," he directed when the MothBots were safely inside.

Brad took a moment to assess the situation by commanding his own MothBot to hover outside the building. He watched the monitors of the other team members as they displayed the Heliorobos flying into their shoebox-type hangars. Now the scene was set for Brad to make his final assault on this dreaded flying enemy.

"Lissy and Bratley, fly your drones into the bank of hangars on the left when I say 'go'. Rena and Slade, you hit the ones on the right. Don't forget to hit your self-destruct switch when you crash into the hangars ... ready ... GO!"

Brad watched the monitors go black, one by one, in rapid succession. At the same time, his ears were confirming the extent of the damage caused by the action. He and the others rushed to the windows where they saw a huge ball of flame rising where the building used to be.

The MothBots crashing into the stacks of Heliorobos had caused each of them to explode with a combined force that blew the building apart and engulfed the remainder in flames.

Brad returned to monitoring his MothBot, happy that it had survived the explosion. He had one more score to settle. He commanded his drone to fly to the compound's training area.

All the suicide bombers had heard the explosion and then seen the flaming aftermath. They stood huddled together as if trying to protect one another from an unseen alien threat.

As Brad's MothBot grew nearer, he saw a figure running toward the group. He could hardly believe his luck. It was none other than Sam Murphy.

In spite of Brad's misgivings about taking the law into his own hands, he could not help himself from exacting his revenge by sending his MothBot full speed into the fleeing figure. His timing was deadly accurate. His target reached the group of suicide bombers just as Brad fired his Lazigun and activated the self-destruct switch.

What had been an attempt to neutralize one threat to Brad's life, turned out to put an entire terrorist organization out of business. The suicide vest of each of them detonated in a chain-reaction explosion that could be heard for miles.

*****

Hector was happy to be home. The adventure in the motor home had been fun, but all that time spent at Lissy's was boring. There were hardly any people around most of the time. It was not that he did not like being alone. It was that his food dish would magically get filled only when someone was around. He would have gladly prepared his own meals if he could have found some way to get outside where all the food was.

Anyway he was home at Brad's place now where he could snuggle in bed next to Brad and Lissy, and together they could watch the birds outside; especially that big red one who thought he was so hot with that pretty face and that crown sticking up like that. Just as Hector was about to make an imaginary leap to capture his prey, he heard Brad speak.

"Lis, I've often wondered if we did the right thing when we decided to attack those terrorists."

"I've thought about that too. Not the part about the destruction of all those Heliorobos, but about the suicide bombers when all their vests exploded at once."

"That was a little tough to hear about," Brad agreed, "but if you think of it another way; they all got their wish to be martyrs."

"Does dying that way really make people martyrs?" Lissy wondered.

"It's clear that the people who do it think so," Brad replied, "because they supposedly will be rewarded with earthly pleasures after they die."

"So, it's a selfish thing," Lissy observed.

"It's beyond selfish to receive training on ways to murder innocent people in order to gain personal rewards," Brad clarified.

"I believe that too," Lissy agreed, "however, I've heard they do the things they do because they want to rid the earth of non-believers and foreigners."

"Murder can never be justified," Brad contended. "Besides that, when a suicide bomber detonates his explosives, he has no way of knowing who the victims will be. He may very well select his intended target, but an explosion kills indiscriminately, and religious folks just like him who believe just as strongly as he does could be killed or wounded."

"Yes," Lissy agreed, "and if he kills a woman who is carrying an unborn child, he couldn't know if that child would be born a believer or a foreigner or anything else. And besides, it just doesn't seem rational that a person would expect to enjoy unlimited pleasures-of-the-flesh in their afterlife when they blow the only flesh they will ever have to smithereens."

"I'm thankful for the fact that this group of terrorists had no more opportunities to cause harm to others," Brad concluded. "I've decided we did ourselves a favor ... as well as the rest of the world ... by doing what we did."

Hector squinted his eyes, kneaded his claws into Brad's shoulder, and began purring vigorously.

PEOPLE, PLACES, AND THINGS

Accomplice = Sells out to terrorists
Arabica = A type of coffee
Arnie and Sludge = Bungling detectives
AutoCAD = Registered Trademark of Autodesk, Inc.
Bell Helicopter = Type of rotary-wing aircraft
Brad Ganderson = Principle character
Calories Ø Not Café = Terrorist target
Captain Ed = Lissy's father
Compound = Terrorist training site
Contact = The Listener's connection to terrorists
Coors = Brad's favorite beer
Culprit = The one who stole Powervert's secrets
DOD = U.S. Department of Defense
David's Place = Powervert's remote research area
Derek Sorsky = Dronomics officer
Detective Nettbron = Medlinton PD
Devouts = Heliorobo operators
Dronomics = MothBot supplier
Fillmore Street = Location of Benjii's warehouse
Frangelico = A liqueur trademark
GPS = Global Positioning System
Gordon, Bratley, and Slade = Powervert scientists
Grubby Ernie's = Lissy's place of work
Hair pretties = Hairpretties4you.com
Hector = Brad's cat
Heliorobo = Clone of MothBot
Hinny = Mahinder al Baq'r from Murphy's past
مقدّس حاربـات = Holy Warriors
iPhone = An Apple trademark

Jason Grilling = A Powervert partner
Lazigun = Powervert's power-beaming device
Lissy Edington = Brad's love interest
Listener = Brad's nemesis
Mahinder al Baq'r = from Murphy's past
Martha = Receptionist at compound
Medlinton = Fictional city in New Jersey
Mohammad = Sahar's new interest
MothBot = Small drone aircraft
Nettbron = Detective
Norman = Deceased founder of Powervert
Pizza Hut = A restaurant chain's trademark
Power-beaming = Sending electricity wirelessly
Powervert = Brad and Rena's company
PPP = Piezo-ignited Poison Pill
Rectenna = Antenna designed to receive electrical power
Rena Offlin = A Powervert partner
Roland Benjii = Murphy's Contact
Saddam = Concert hall suicide bomber
Sahar = Alias Sam Murphy
Sam Murphy = Brad's archenemy
Santa Margarita = A wine producer's trademark
Scientific Atlantic = Manufacturer's trademark
Slade, Gordon, and Bratley = Powervert scientists
Spooks = Extremists inside compound
Triton 1500 = A Magellan GPS
UAV = Unmanned Airborne Vehicle
Warehouse on Fillmore Street = Contacts HQ